Good Morning, JUDY!

Good Morning, Judy!

Jeanine
Steuck

AUGSBURG Publishing House • Minneapolis

GOOD MORNING, JUDY!

MANUFACTURED IN THE UNITED STATES OF AMERICA

1

Judy Steuck was a healthy, effervescent 22-year-old graduate student from the University of Wisconsin, Madison, traveling in Spain. One night in December 1975 she went to sleep in a hostel in Madrid. She was not to awaken until the following February. After nearly eight weeks in a coma and on the brink of death, Judy astonished her nurse with the simple greeting: "Good morning."

I had traveled across the Atlantic and back, hoping and praying and waiting to hear those two incredible words.

Judy had completed a semester of graduate study in Europe with a group of students from the Waisman Center in Madison, Wisconsin, and had spent five happy, carefree days in Austria with a girl friend who had married an Austrian. Before meeting her homeland-bound group in Amsterdam, Judy traveled to Spain to spend a few days with a childhood girl friend, Judith Ann Warner, who had been studying in

Madrid with a group from Beloit (Wisconsin) College. Judy decided to stay in a hostel across the street from the Beloit College group.

After an evening together at a Spanish opera, Judy and Judith Ann returned at midnight to Judy's hostel. They asked the woman who operated the hostel for a heater to take the chill off Judy's room.

Before the woman left, Judith Ann asked, in Spanish, "Will it be safe to leave the heater burning through the night?"

"No, it won't be safe," answered the woman, "because of the fumes that could be breathed in."

"Will you show us how to turn it off?" Judith Ann asked.

"I'd rather do it myself," the woman answered, "because I want to be sure it is done right. I'll be back in half an hour at 12:30 and I'll move the heater into the hallway out of your friend's room."

Judith Ann told Judy what had been said and the girls made plans to meet the next day at 2:00. Judith Ann could not meet her until afternoon because in the morning she had to make preparations for her job. She planned to stay in Madrid to teach English. They said good-night.

At home in Wisconsin, December 17, 1975, I arrived home from work for lunch and checked the mailbox for a letter from Judy. As I came into the house, the phone was ringing. It was my husband, Gene. He sobbed, "Something terrible has happened."

I knew intuitively it must be Judy. I said, "Get hold of yourself and tell me what's happened."

He continued to sob and said, "Judith Ann's mother, Lou, just called me and said our Judy is in a deep coma in Madrid, but they don't know what's causing it. We are to get a doctor to come to our house by 4:00 so he can talk to the doctors in Spain. Lou will be coming over with the name of the hospital and their number. In the meantime, we're supposed to try to think of anything we know that may have caused the coma. I'm going to come home as quickly as I can."

I was stunned. I had felt a premonition for about a month that something was going to happen to Judy, but I hadn't expected anything like this.

Dear God, our Judy was in a coma! What could have caused it? She was always so healthy.

I took a bite of the salad I had prepared for my lunch, but it stuck in my throat. My stomach felt like lead. I put my food aside and made arrangements for a doctor to come to the house, called work and told them our sad news, and cancelled some appointments.

Gene came home and we went to church and talked with our pastor, Glenn Nycklemoe. He offered to do anything he could to help. We prayed for Judy and left. I found out later a prayer chain was started almost as soon as we had gone. At home we waited and prayed. I was confident God would help us find the strength to do whatever would have to be done.

Lou Warner brought us the name of the hospital and the number we were to call. The minutes inched by. Finally 4:00 arrived and the doctor came. With him was a medical student who spoke Spanish.

Shortly after our doctor arrived our 15-year-old daughter Janet came home from school. I went to the door to meet her. She came strolling in and said, "Who's here?"

"Janet, honey, Judy is very sick and she's in a coma."

Janet shouted, "What?" and started to cry. Her Sunday school class had discussed the Karen Quinlan case the Sunday before and each person had given an opinion on "pulling the plug." Karen Quinlan came to her mind when I said "coma."

I wanted to be with Janet but it was time to make our call to Spain. She went upstairs to her room in tears.

The medical student placed the call to Madrid, and we learned the horrible news. Judy had carbon monoxide poisoning from a faulty heater. The Spanish doctors wanted someone to come to Spain immediately.

After the call was completed Janet came downstairs, trying to act composed. Our doctor sat down with her and explained what happens to a person who has gone into a coma from carbon monoxide poisoning, describing how serious cases became rigid. But, the doctor added, "Judy is not rigid." Janet was very relieved.

Who should go to Spain, and when, and how? We decided I should go, because I had a passport that could be renewed, but with whom? Gene had never been on a plane. And what about flight reservations? It was December 17, eight days before Christmas, and two airlines were on strike.

All of a sudden everyone was gone. Gene and I were left to decide how to make arrangements. We felt helpless and alone.

I called my brother Harold Cole, a businessman in Janesville, Wisconsin, and said, "I need help." I explained the situation. Harold and his wife Jeanne came immediately. He could see our bewildered state of mind and took command of the situation. He managed to make reservations for the *last two* seats on a flight out of Milwaukee at 2:00 the next afternoon. This was the first of many small miracles. God's hand was already at work.

After much discussion Harold said he thought he should go with me (thank God), since he had a valid passport and had been in Madrid.

After all arrangements had been made, I called our son

Jerry, who had just come home from work. He could not believe that something so terrible could have happened to Judy. I tried to calm him down and told him to come over to spend the night. He agreed to come over early the next day. Then I knew it was time to call Jim.

Judy had met Jim a few days before she came home after graduating from the University. They were attracted to each other immediately and saw each other whenever they could during the summer. Their romance blossomed and Judy confided in me before she left for Europe that she and Jim had talked about making wedding plans after she returned from Europe.

When Jim received my call he didn't seem too concerned at first, because that very day he had received a telegram from Judy wishing him good luck on his exams for his master's. He found it hard to believe that Judy was in a coma. I asked him to call the Waisman Center to see if they would get word to the group in Europe so they would know what had happened to Judy.

Before I hung up I said, "Pray for her, Jim."

He sadly replied, "I do, every night."

2

We were on our way at 6:00 the following morning. The impossible had happened.

There were more small miracles the next day as we rushed to Chicago to get my passport renewed. Time was precious.

I was the only one at the small photo shop to get my passport picture taken. The photographer said, "I don't understand this. We usually have a big line waiting."

Then to the passport office by 9:15. We explained the situation and the man said to come back in an hour. Harold had expected a wait of at least five hours.

Next, to the bank for traveler's checks. (Harold had lent me some money.) There was no line at the bank until we finished; then there were lines everywhere.

After confirming our flight reservations, we went back to the passport office, where my passport was already waiting.

We were driving back to Milwaukee for takeoff by 10:30 A.M. Unbelievable!

We made the 2 P.M. departure easily. Two apprehensive travelers arrived in Madrid at 7:30 Friday morning, Madrid time, and were met by Judith Ann and her friend Lisa. From Judith Ann we learned of the nightmarish events that had occurred the night before:

After leaving Judy, the woman at the hostel became engrossed in a TV program and forgot to return to Judy's room to turn off the heater.

Six months later, on May 22, 1976, while Judy was lying in her bed, she heard the humming of the fan in the window and remembered:

At 12:30 I looked down the hall to see if I could see the woman anywhere. Instead, I saw a man walking down the hall toward my room. I became frightened and locked the door and left the key in the lock as an added precaution. I was very tired and hot and lay down to wait for the landlady to come back.

At 5:00 A.M. the man in the room next to Judy's heard groans and movements in the bed. He wasn't concerned because he thought there was a married couple next door.

One year later, on December 12, 1976, as Judy lay

in bed in a basement bedroom next to a furnace, which was switching on and off, she remembered:

I tried to get up to turn off the heater and I tried throwing back the covers. I was so weak I could hardly move. I felt like I was moving in slow motion. I fell back on my bed and called, "Mom, Mom—help me."

At 11:30 A.M. the landlady noticed the door closed on Judy's room. She knocked and got no answer. She knocked the key out of the door with a pair of scissors and unlocked the door with her key. She found Judy unconscious. Confused and excited, she grabbed the heater, turned it off, and took it out of the room. Eventually she called a doctor.

A very young doctor arrived on the scene. Because the bed was in disarray, he became very excited and thought Judy might have been raped (a very serious crime in Spain) or might have taken an overdose of drugs or alcohol.

Fourteen months later, on February 20, 1977, as Judy was in church listening to a sermon called "Life after Life," she recalled:

A man and woman were speaking in Spanish, very excitedly, and the woman was crying. Then I saw a very bright light.

The Spanish woman finally called an ambulance. Then she called Judith Ann, who rushed over with her friend Lisa. She noticed the heater gone and asked the landlady if she had turned the heater off. She said she did, so Judith Ann dismissed the idea of gas poisoning.

By this time Judy's eyes were rolled back in her head. She was very cold because her circulation was almost nil. Judith Ann kept shouting to the doctor, "Is she going to die? Is she going to die?"

The doctor would not answer.

He put a hot water bottle on Judy's stomach. (Because of the variance in temperature between Judy's body and the hot water bottle, third degree burns resulted.) He asked Judith Ann to help him turn Judy over so he could give her a shot to stimulate her heart, which had almost stopped.

Judy finally arrived at the hospital at 3:00 P.M. Judith Ann saw the landlady at the hospital and asked her again about the heater. She finally admitted she had turned the heater off *after* she found Judy. Precious hours had been wasted.

Judith Ann told the doctors about the heater. When they learned what they were fighting, they gave Judy steroids to prevent cerebral edema (swelling of the brain). Fifteen hours had elapsed since the fateful heater had been brought to Judy's room.

After Judith Ann described all these horrible events, I thought, "How can Judy still be alive? God must have some special reason for letting her live."

3

We went to Judith Ann's hostel and met her teacher, Renee Fendrich, who spoke Spanish fluently. She went with us to the Hospital of the Red Cross, where Judy was now a patient, to interpret for us. The head neurologist told us that Judy was in a deep coma and that he had found evidence of brain damage. "The best thing would be to take her home as quickly as possible," he said. "But if she shows any signs of coming out of her coma, she should not be moved."

We were then allowed into the intensive care unit to see Judy. When I'd last seen her, three and a half months before, she was happy and vibrant. Now she lay deathly still with only a sheet over her body. She had an IV (intravenous tube) in a vein in her shoulder, and she was receiving oxygen.

"Judy," I called softly, then imploringly, "Judy! It's

Mom. . . ." There was no response. Would she ever be able to answer me? Realizing we could do nothing to help her there, we left and immediately went into action to move her home.

Harold called our doctor in America and explained that the Spanish doctor wanted us to make arrangements to bring Judy home as soon as possible. I told Harold to ask if it would be better to take Judy directly to the University Hospital in Madison. I wanted her to receive the best care. The doctor assured us that the Beloit hospital had all the latest equipment and a neurologist whom he could consult. Harold asked the doctor to relay our message to my family.

We then went directly to the TWA office and told the reservation clerk about our dilemma. She said, "Before I make any reservations, you need to have a fitness-for-travel permit form filled out by the doctor at the Hospital of the Red Cross. Then it will have to be wired to Kansas City for approval before I can check on reservations. I'm sorry." We thanked her and rushed to the hospital with the form.

We took a taxi to the U.S. embassy and found it closed until 2:00 P.M. We did not want to waste any precious time, so we took a taxi back to the hospital to see if the permit was ready. Because the form had to be translated into English, it was not ready, so we decided to go back to the embassy.

We tried for more than an hour to hail a cab, but they all went right past us, even though they carried no passengers. This seemed strange because we had not had trouble getting cabs before.

Harold and Judith Ann stood on one corner and Renee and I on another. We wanted to make sure we had someone at each corner who spoke Spanish, but no cabs stopped.

In desperation Judith Ann stopped a bus and asked if any buses went by the embassy. She received a flat no and the door slammed in her face. What were we going to do? The embassy closed for the weekend at 5:00. We knew we would need their help and we also had to notify them that we were in Madrid to take care of Judy.

We went back to the hospital and asked the receptionist if she knew why the cabs were not stopping, but she had no idea, nor did anyone else. She told us which bus would take us within six blocks of the embassy. It was 4:00. We were racing against time.

We got off the bus and practically ran all six blocks. The last block Harold took my hand and almost dragged me, panting, up the long, steep hill.

We arrived in time and explained the horrible details to the consulate's secretary. She took us right to Joseph Cheevers, first secretary of the embassy. Again we repeated our story. "And she's still alive?" he asked.

He was upset because he had not been told about Judy earlier. Any misfortunes which occur to American citizens are supposed to be reported to him immediately. But the landlady did not call the police or inform anyone about Judy, probably because she feared her hostel would be closed.

We asked him if he thought the hostel would have liability insurance. He said, "I doubt it very much. Spanish laws regarding liability are just the opposite of American laws. We just had a case where the claimant had slipped on a rug in a very nice hotel and was seriously injured. The verdict came back—she should have watched where she was going. But I will talk with the landlady and call you later this evening."

We asked if he knew any reason we could not get a cab.

16

He did not. We thanked him and left. As we were leaving, I looked at the clock. It was 5:00!

By asking directions we managed to take the right Metro back to the hospital. We picked up the permit and at 7:00 P.M. went to Judith Ann's hostel, where she had reserved a room for us. The lobby looked so gloomy and foreboding that Harold said he did not want to stay there for the night. He made reservations at the Hotel Melia, a beautiful hotel where he had once stayed. As we were leaving we heard that cab drivers had gone on strike without telling anyone. We had no way to transport our luggage, so we had to stay at the hostel.

Judith Ann left us for much needed rest. Harold and I went to our room at the end of a long, dark hallway. After we put down our luggage, Harold said, "I'd better walk down to the TWA office with the travel permit so they can start processing it. It's only a few blocks."

I was alone for the first time since we'd arrived in Madrid. Our room was cold and gloomy. I was shivering, even with my coat on. Suddenly I thought of Judy asking for a heater and going into a coma. I was struck by the impact of what had seemed like a terrible nightmare. In complete desolation and despair, I felt as though my whole world was crashing down around me. Our dear, happy Judy, who had so many plans for the future, would probably never fulfill any of them. I sat on the bed in a stupor, unable to cry.

Then I remembered that it was December 18, the night my church circle had its Christmas party, an event to which I always looked forward. I knew the women would pray for us. "Maybe they're praying right now," I thought. This gave me strength.

Harold came back and suggested we find a restaurant, since we had not eaten since morning. We were directed to a place nearby. Though we didn't know what we were ordering, the food was good. We went back to the hostel to try to get some rest.

I drifted off in a fitful sleep and was awakened suddenly by the shrill ringing of the phone. All I could think of was Judy. I wondered what time it was and who could be calling. Groping for the phone, I found it on the wall near the bed.

It was Mr. Cheevers. "I went to see the landlady and it was just as I had expected," he said. "She has no insurance of any kind. I'm sorry. I would like to do anything I can to help. I won't be in my office tomorrow but if there is anything I can do for you, call the embassy. I will leave the number where I can be reached."

I thanked him and replaced the receiver and looked at my watch. It was 11:00 P.M. People in Spain have very late evening meals and never go to bed until after midnight, so late calls are not unusual.

After a quick breakfast the next morning, Judith Ann went with us to the hospital to act as our interpreter. We wanted to see Judy again and to find out if there was any news about our permit.

We were told to go right into the office of Dr. Estaban. A young man was sitting with the doctor, and he stayed there as we came in and sat down. The doctor and Judith Ann exchanged some fast Spanish conversation. We were then informed that the fitness-for-travel permit had been denied by Kansas City. They required that a doctor accompany Judy on the flight home and the assurance that there would be no possibility of Judy's death in flight.

4

The young man in the room with Dr. Estaban was Dr. Luis Landin, who had volunteered to accompany us to America, even though it meant he might not be home to spend Christmas with his young wife and baby.

Dr. Luis had a valid passport but no visa. How could he get one now? It was Saturday. Harold called the number Mr. Cheevers had given me and explained our problem. Mr. Cheevers offered to meet Dr. Luis and Harold and issue the visa himself. Harold thanked him and said they would leave immediately.

But the cab strike was still on. The doctor called the cab company and asked if we could get a special cab because of the emergency, but they said it was impossible. Dr. Estaban then gave Dr. Luis permission to be away as long as necessary and told Dr. Luis to use his car to go to the embassy, which was many miles from the hospital.

Before we left, Harold and I went into the intensive care unit. Judy was motionless, her eyes closed. She was receiving oxygen. There was no change.

Harold and Dr. Luis left for the embassy, and it was up to Judith Ann and me to go back to TWA for another permit. Then we had to take the permit to the hospital, have them fill it out correctly, and return it to the airline office—all before they closed at noon. We had to take the Metro. Even though we weren't sure where we were going, we managed to get the permit. We dashed back to the Metro. It was already 11:30 as we got off. As Judith Ann took my hand and dragged me across the street, she forgot herself and began talking in Spanish. Then she said desperately, "I'll run the rest of the way to the hospital. We must get the permit there as quickly as we can. I'll meet you there."

"O.K.," I said breathlessly. As I followed at a slower pace, I thought, "There is absolutely no way we can get the permit filled out and back to TWA by noon." I forgot momentarily how God had been intervening for us all along.

When I arrived at the hospital, Harold and Dr. Luis had already returned with the doctor's visa. Dr. Luis called TWA and told them we couldn't possibly get our completed permit back to them before closing. They said they would stay open until we arrived.

The new permit said: "Diagnosis: Neurologic coma. Intoxication by carbon monoxide. Severity & Immediate Prognosis: Unresponsive due to deep coma. Late prognosis poor. *No immediate danger of death*. The passenger will be accompanied by a doctor."

Dr. Luis rushed us in his car to TWA and Harold dashed in with the permit.

When we took Judith Ann to her hostel, Harold asked the doctor (with Judith Ann interpreting) if he would mind taking our luggage to the Hotel Melia. Dr. Luis said he would be glad to, but he didn't know if his tiny Fiat would hold all our luggage and three passengers. We all managed to squeeze in.

Dr. Luis had to go to the bank before 2:00 to get money for his trip. We exchanged some of our traveler's checks for Spanish currency.

We finally arrived at the Hotel Melia. Dr. Luis had been gone from the hospital for more than three hours and had driven his own car many miles. Harold wanted to pay him for his time and the use of his car, but he would not accept anything. We checked in and were shown to a big, bright, airy, luxurious room. My spirits lifted at once.

Judith Ann and her friend Lisa came to our hotel a short time later. Judith Ann said, "We'd better go and get Judy's luggage and her personal belongings from the hostel. The landlady said she would really like to meet you and tell you how sorry she is, but she and her husband are very emotional."

I knew I couldn't take an emotional scene. I couldn't take the chance of breaking down. Too much had to be done. And I did not want the landlady's face to haunt me all my life if Judy didn't recover. I asked the three of them to go without me.

I was alone again, but this time it didn't seem so depressing in a bright and cheery room. I took my Bible out of my bag and I read the 23rd Psalm. Then I read on:

> Who may climb the mountain of the Lord and enter where he lives? Who may stand before the Lord? Only those with pure hands and hearts, who do not practice dishonesty and lying. They will receive God's own goodness as their blessing from him, planted in their lives by God himself, their Savior. These are the ones who are allowed to stand before the Lord and worship the God of Jacob (Ps. 24:3-6 LB).

I felt a great sense of peace. To me this description of a

pure person fit Judy. I knew God would keep her wrapped in his loving arms and no matter what happened he would always take care of her. Gentle tears dropped on my pillow.

Judith Ann, Lisa, and Harold returned with Judy's luggage. Seeing all Judy's personal things gave me an empty feeling. It seemed she should be right there with them.

I knew Judy had planned to meet two young women from her group in San Sabastian but I did not know who or where. I found her journal in her tote bag and checked for an entry that might tell me. There was nothing. We had no way to let them know what had happened to Judy.

With Judy's luggage were many gifts she had bought for our Christmas. I put them aside, praying some day she would be able to give them to us herself.

I found the last letter I had written to Judy, and I read it over. The last line said, "Judy, I wish you wouldn't go to Spain."

That evening Harold took Judith Ann, Lisa, and me to the hotel's elegant dining room for an American meal. The chairs were a deep blue plush velvet and the waiters were eager to respond to our slightest wish. Harold was trying to do everything he could to make this ordeal more bearable. But I couldn't eat. Judith Ann, who had almost lost her voice from a cold, whispered, "What a time not to be hungry." She couldn't eat either.

Before we retired for the evening, Harold called Mr. Cheevers and asked if there was any way we could get the U.S. Air Force stationed in Madrid to fly Judy home if TWA should turn us down. My husband Gene is a veteran and Harold thought they might do it for a veteran's daughter. Mr. Cheevers said he would check, but it would probably cost as much as the airlines. I was much more concerned about getting Judy home than about the cost. I told

Harold we would just have to get a second mortgage on our house.

At 9:00 Sunday morning, December 21, the reservation clerk called to tell us that TWA had accepted Judy as a passenger. Though the office had closed at noon Saturday, she had worked Saturday afternoon and Sunday morning to make our travel arrangements. We were to go to the airport to confirm our reservations. Another small miracle in answer to many prayers.

We called the desk to see about transportation to the airport. The cab strike was over.

At the airport many kind and sympathetic people made arrangements for special accommodations. We needed seven seats and the only space available was in the first class section.

Harold was very concerned about paying for our fares. He said, "If the airlines won't accept my American Express Card, which has only a $500 limit, I'll have to call my bank so they can verify I have sufficient funds to cover the check. I don't know what else we can do."

We went with Judith Ann and Lisa to see Judy and make arrangements for our departure the next day. Judy was the same. She showed no sign of coming out of her coma. The doctors agreed that she could be moved. I breathed a sigh of relief.

Back in our hotel we called the family to tell them the good news. Judy would be home for Christmas!

Harold said he wanted to take Renee and Roger Fendrich, the teachers in charge of the Beloit College group, and Judith Ann and Lisa out for dinner. We went to a beautiful restaurant, with a breathtaking view of Madrid, for a delightful meal. Everyone was able to eat this time.

We all agreed to meet at Harold's pool in the summer

and have a party for Judy and Jim. It was our way of saying, "Judy will recover."

I hated the thought of Judith Ann and Lisa staying in the hostel, where they had moved the night before, so I asked them to stay with us. They jumped at the chance. We brought their luggage to our large room, where we spread mattresses on the floor. Everyone had a fairly good night's sleep, knowing all arrangements had been made to bring Judy home.

5

Monday morning, December 22   After breakfast we took a cab to a department store. We had to be back at our hotel by 10:00 because the hospital was going to call to tell us the amount of our bill, but I wanted to buy something Spanish for Judy so I could show her when she recovered that I never doubted she would get better.

We got back just in time for the call from the hospital. Our bill amounted to $850. We would not have enough money to pay Dr. Luis. Harold called Gene and told him to get $500 from our bank to have ready to pay the doctor.

Judith Ann and Harold went to the bank to exchange our remaining traveler's checks into Spanish currency so we could pay the hospital. Lisa told them from experience, "It takes at least an hour to get waited on. The lines are so long and so slow. You'll never get back in time to get to the hospital by 11:00."

Our flight was scheduled to depart at 1:00 P.M., but many preparations had to be made at the airport. The airlines wanted us to leave the hospital by ambulance at 11:00.

Time again was precious. At the bank Judith Ann and Harold received their number and went to the end of a very long line to wait to be called. Suddenly their number was called. Amazed, they went to the front of the line and they were waited on immediately. God's hand was again at work.

We arrived at the hospital at exactly 11:00, but we had to wait for the staff to complete some paperwork. The ambulance was so small there was room only for Judy and Dr. Luis. We followed by cab.

Dr. Luis stayed with Judy while Harold and I checked in with the reservation clerk. Arrangements were made for special oxygen on each flight we were to take. Then Harold asked the big question: "Will you accept my American Express Card, even though it has a $500 limit?"

The clerk said, "Just a minute, I'll check." In a short time he was back. "That will be just fine." Everything was ready for our departure.

Judith Ann stayed until we were ready to board the plane. As I turned to go she called, "I love you." The dam holding back my tears almost broke.

We boarded the plane and were led to the first two seats in the first class section. Judy had been placed across from us in a type of hammock hung from the ceiling. A curtain was pulled around her for privacy. Four seats were folded down under her and special oxygen tanks were placed under them. She was catheterized and had an IV in her shoulder. An intubation tube had been placed in her throat for breathing. (It probably had been this tube, pushed quickly into Judy's mouth in Spain to save her life, that had knocked the cap off her front tooth.)

Dr. Luis sat right next to Judy during the entire trip and was so attentive that the flight attendant hesitated to distract him long enough to ask if he wanted anything to eat. He used a very large rubber syringe to clean out Judy's throat when she made gurgling sounds. We were unaware at the time that she had contracted pneumonia.

The pilot expressed his sympathy and said he would do everything he could to make the flight smooth. It was sleeting and very icy when we arrived at Kennedy Airport. As the pilot put the plane down—light as a feather—everyone on the plane applauded. When I thanked him, he said, "I'll pray for her recovery."

We were met at the airport by an ambulance and transported to LaGuardia Airport, where we had a three-hour layover. TWA invited us to use their VIP room and to use their Watts line for calls.

Harold called our doctor in Beloit to confirm all plans made for Judy's arrival at O'Hare. He then called his wife Jeanne and told her Dr. Luis would be their guest for the night. Dr. Luis had been very nervous about coming to America because he did not speak English. Harold invited him to stay all night at his home in Janesville.

I called Gene to let him know we were almost home. He said the family was anxiously preparing to go to O'Hare to meet us.

Dr. Luis hadn't had any sleep for 27 hours. We told him to rest on the couch, and he fell asleep immediately. One of the ambulance attendants cared for Judy and watched her vital signs as the doctor rested.

We arrived at O'Hare with our precious cargo at 9:30 P.M., December 22. As I stood by Judy waiting for all the passengers to leave, familiar faces started to appear—two paramedics from Beloit, whom I knew personally, Brian Brown and Jay Kurth, came rushing in to get Judy, followed by our doctor. Then Gene came in. He grabbed me and held me tight. It had been hard for him to be home with nothing to do but wait.

Judy's group from the Waisman Center had landed at O'Hare two hours earlier, unaware of Judy's misfortune. Judy's main luggage had been left back in Amsterdam. The Waisman Center had mistakenly sent word about Judy to O'Hare instead of to Amsterdam. We had been paged at O'Hare to tell us Judy was not on their flight at the same time we were in flight bringing her home.

The paramedics took Judy to the waiting ambulance, where a medical student who spoke Spanish interpreted as Dr. Luis informed our doctor of Judy's condition.

Our daughter Janet and our son Jerry were anxiously waiting for us with a family friend, Deane Leavitt, who had driven them to O'Hare.

Janet said Pastor Glenn Nycklemoe had already started a fund in Beloit to help defray the mounting expenses. The Fire Department Union donated the ambulance run to O'Hare, and the paramedics donated their time. People of all faiths in Beloit and neighboring towns had been praying for Judy's safe return and recovery. I was overwhelmed. Ours was truly a caring community.

It was apparent to me that the small miracles which had taken place since Judy's accident were the result of the fellowship of prayers that had supported us since Judy's accident. I thanked God for such a blessing.

At the hospital, a crew was waiting at the emergency entrance for Judy's arrival. We were surprised to find the ambulance had not arrived.

Before leaving O'Hare the paramedics suctioned Judy's lungs and got up a great deal of mucus. On the trip to Beloit they stopped six times and turned off the flashing red lights and motor so they could hear more clearly to take her vital signs.

The ambulance finally arrived about an hour later. They rushed Judy into the hospital and everyone who had been waiting for her arrival dashed into action.

A short time later our Beloit doctor told us, "It's a good thing you got Judy home when you did. Praise the Lord she made the trip. But Jeanine, why didn't you tell me about her burns?"

I said, "We were told the burns were not serious and we didn't have to worry about them."

The doctor exclaimed, "She was burned so badly she has a round circle just like a brand on her stomach, probably from the ring on the hot water bottle. The burns on her stomach are seriously infected. She has a high fever and pneumonia. We will need your permission to do a tracheotomy right away in the morning. And we'll do an EEG— a brain scan—as soon as we can."

We signed the paper for the tracheotomy and also signed a paper giving the paramedics permission to observe Judy's condition daily. It was to be a learning experience for them.

Judy was placed in the intensive care unit (ICU) on a Stryker frame. She was near death. Her name was immediately put on the critical list.

We were given the ICU number and told we could call any time of the day or night to check on Judy's condition.

My parents were at our house waiting for our return. They had come back to Beloit from Florida to spend Christmas with us, unaware of Judy's horrible accident. They were not told about Judy until Janet and Jerry and a family friend, Arnold Lee, met them at O'Hare. We wanted to make their trip more bearable.

I hoped there would not be an emotional scene, but I

needn't have worried. Everyone was trying to be brave. My mother put her arms around me and hugged me. When she said I felt like a piece of steel, I wasn't surprised. I had steeled myself since Judy's accident and would not allow myself to let go.

6

In the morning we called ICU. Judy was as good as could be expected after her tracheotomy. We ate a light breakfast. Then people started arriving.

Harold came with Dr. Luis. We paid him and gave him a pen and pencil set as a gift. I also wanted to give his baby daughter something. I thought of my favorite Christmas tree ornament, a sequined rocking horse Gene's Aunt Rose had made for us. I took it off the tree and handed it to Dr. Luis. He seemed very pleased.

Jim arrived, and we introduced him to Dr. Luis as Judy's boy friend. Then our son Jerry said something to the doctor in Spanish. I said, "I didn't know you spoke Spanish, Jerry."

The doctor said, haltingly, "He speaks Spanish like I speak English. Not good."

Jerry took a picture of Judy out of his wallet and gave it

to the doctor. He was happy to get it and put it in his wallet.

Lou Warner, Marcia Lee, and Darlene Nelson, a Spanish teacher at the high school, came to take Dr. Luis back to O'Hare. I kissed him goodbye and told him how grateful we were for all he had done for us.

Our family and Jim went to the hospital, but only Gene and I were allowed to see Judy. Jim wanted to see her, but the doctor did not think it would be wise for him to see her yet.

Jim looked lonely and forlorn. He had brought a beautiful red azalea plant with him and a big card. Flowers were not allowed in ICU so he brought the plant back to our house and took the card back home with him. He said he would be back on Thursday—Christmas Day.

Gene and I went to church to pray and to talk with Pastor Nycklemoe. We thanked him for all he was doing to help, which included acting as our spokesman for the news media.

Gene told me he did not think he could have endured the wait while I was in Spain without Janet. She was a great source of strength to him. Janet had strong faith that Judy would recover. Whenever Gene became depressed, Janet was there to offer encouragement and support.

Janet carried out all commitments she had scheduled, including playing a difficult part in a clarinet quartet for the Christmas band concert December 19. Her director told her she could be excused, but she said she would do it. Gene went to the concert and was proud of her flawless performance.

In the afternoon I went to my bedroom to rest. As I lay there, unable to sleep, I heard the phone ring. The ICU nurse had said she would call if there was any change in Judy's condition. I wondered if she was calling. I listened

intently and heard Gene answer the phone and talk in muf-
fled tones. He hung up and I thought I heard him say, "It's
Judy—it's all over."

My heart sank. I thought, "Oh, no, dear God, you really
did want Judy in heaven with you after all." As I waited
for Gene to come in, dreading what he would have to tell
me, I prayed for strength to accept Judy's death. But he did
not come in. When I could not stand the suspense any
longer I got up and went into the living room and asked,
as calmly as I could, "Who was on the phone?"

"Oh, it was Judy Miller," Gene nonchalantly replied.
"She's coming over.

I was so relieved. Judy was still alive! I thanked God and
sat down in a chair, trying to realize Judy had not died.
When Judy Miller came over with her husband Hans, I
was still dazed and not able to comprehend much of the
conversation.

December 24   Judy's arms were pulled up next to her
body in a rigid prenatal position. Her hands were twisted,
and she had burns on both arms from the hot water bottle.
The opening in her throat had steam blowing on it to keep
it moist. All her vital signs were being monitored. An IV
was now in her arm instead of her shoulder. A catheter
attached to tubing emptied her urine into a plastic bag at
the side of the Stryker frame. On the frame Judy seemed to
be in a canvas sandwich held tightly by straps.

"Our poor darling!" I sighed. "Mom and Dad are here
with you now, Judy!"

She lay motionless with her eyes closed. Suddenly she
had a violent body tremor. I was shocked and asked the
nurse why. She said she would tell me later.

After we left Judy's room the nurse said, "It is not known

if coma patients are aware of what is going on about them, but it would be best if you did not talk about anything you are concerned about in front of Judy."

I have always been grateful to that nurse. None of us knew anything about comas. We had everything to learn. After that we made only encouraging remarks in Judy's presence.

In the evening after going to the hospital to see Judy, we went to our church for the special Christmas Eve service celebrating the miracle of Christ's birth. All the choirs sang Christmas hymns. As Janet's choir, the Disciples of Distinction, got up in front of the congregation, I felt a great void. On her last card home Judy had said how much she was looking forward to being with us Christmas Eve and hearing Janet's choir. I prayed for God to strengthen my faith and give me the courage to endure whatever was ahead.

As we were leaving the church I saw anguish in many eyes, but no one seemed to know what to say. Then a friend came up to me, put her arms around me, and said, "We love you." That simple gesture said it all.

December 25   After Christmas breakfast with my parents, we distributed our gifts. Judy's gifts were placed back under the tree. As we opened our presents, we tried in vain to show enthusiasm.

Judy had sent five large boxes home from her travels in Europe. She had said, in one of her letters, that we were not to open any of the packages because most of them were gifts. I had put them all in her room to await her return. Later I put our Christmas gifts for her with the packages she had sent. Would Judy ever be able to open her gifts? I prayed that she would.

Many wonderful friends and neighbors had brought casseroles, salads, pies, cookies, and other goodies. Their gifts made our Christmas dinner.

Jim arrived just as we were going to have dessert.

"Would you like to have a piece of pie with us?" I asked.

"Yes," he answered. "I'll help you serve it."

In the kitchen he put his arms around me and said, "I think I can help bring Judy out of her coma."

I didn't know Jim well, but I could see how sad and lonely he was. I said, "I know you can, Jim. Judy told me you and she were hoping to be married. I know she cares about you very much."

He seemed relieved to know that I was aware he and Judy were serious about each other.

As Jim and I left for the hospital, he said, "I'm glad I didn't see Judy when I came before. I've had time to get myself psyched up before seeing her. I've never been in a hospital much before, but I've watched some hospital scenes on TV so I have some idea of what to expect."

For most patients only one visitor is allowed for five minutes each hour. Because of Judy's special circumstances, two were allowed in her room as long as they wanted.

Jim hadn't seen Judy for three and a half months. He took the shock of seeing her condition well. He bent over her and said, "Hi, Judy. It's Jim. How are you?"

There was no response. I felt sorry for him.

He said, "It's so good to see you again, honey. I'm glad we're together again."

Judy just lay with her eyes closed and her arms pulled up tight against her chest. Jim kept talking to her and touching her.

December 26   Jerry had stayed at our house while I was in Spain so he could be with Gene and Janet. He slept in

35

Judy's room. Now he came to me with tears in his eyes and said, "I just can't stand to sleep in Judy's bedroom with all her Christmas presents and those boxes she sent home from Europe. It's too much for me to take."

I felt sorry for our son. "You don't have to sleep in her room if it bothers you," I told him. "Why don't you just sleep on the couch in the den?"

He was relieved. Later he came into the kitchen as I was preparing our meal and said sadly, "You know, when Judy was in Europe I sent her a letter and told her all my frustrations and doubts about everything that had been bothering me for a long time. She wrote back a nice letter of encouragement. Now it seems like a last confession to her."

He broke into sobs and said, "She is the closest thing to a saint there could possibly be on this earth. Everyone liked her, and she liked everyone she knew. I never heard her say anything bad about anyone. She was almost perfect. And now this had to happen to her. Why couldn't it have happened to me instead? She's always had so much more going for her than I have."

I felt deep compassion for Jerry. I'd always known he had special affection for Judy, but I hadn't realized the depth of his feelings.

I said, "Jerry, I'm sure Judy wouldn't want you to feel that way. We can't question why these things happen and why they happen to someone who is good and seems to have so much going for her. Maybe God has some special plan for her. Why don't you try to do everything you can to make her proud of you? If you find you're doing something you don't think she would like, think of what she told you in her letter and do what would make her proud."

A few days later Jerry started to sleep at his own apartment again, but he continued to eat evening meals and Sunday dinners with us.

December 27   Judy's fever remained high—104°. No medication would counteract the infection in the burns on her stomach. An electroencephalogram (EEG) charting the electrical impulses generated by her brain showed abnormal tracings, indicating brain damage. She remained in critical condition.

December 29   Judy's fever was still high. The infection was from a Spanish germ in the infected burn on her stomach. No medication helped. She was put in isolation. The doctors did not want the germ, which could not be controlled, to spread anywhere else in the hospital.

Before entering her room we had to put on white gowns, wash our hands in an iodine solution, and put on rubber gloves. When we left we had to take off our gowns, put them into a bag, take off our gloves, and wash our hands again in the iodine solution.

We could touch Judy only with gloves on. Dear God, we were not even allowed to kiss her! We told her we did not want to infect her.

The days and nights dragged on. Almost two weeks had elapsed since the fateful call.

Jim was on semester break. He came to see Judy every night, talking to her all the time he was with her. He got no response. One night as we pulled into our driveway after spending the evening with Judy, I sensed his frustration and said, "Jim, would you like to pray with me?"

"Yes," he answered desperately. He took my hand and we prayed for Judy. Sharing our burden with God made both of us feel better.

December 30   Jim and I went to see Judy. Her eyes stared into nothingness. Her arms were pulled up against her chest. The nurses had tried wrapping her arms in gauze to keep them down at her sides, but she pulled them out and had them back up to her chest. Jim kept talking to her during our visit, without response.

Outside her room I said to Jim, "Janet told me there is a beautiful meditation room on the first floor. Would you like to go there with me?"

"Yes, let's go there now."

We found the beautiful chapel all in red with very subdued lighting. There was a round altar with a kneeling rail. We went in and prayed. Jim took my hand and held it tight.

I said, "We just can't question why, Jim."

"Well, I do," he answered sorrowfully.

"But God can cause all things to work together for good," I said. "Maybe some day Judy will be able to help other people."

Jim did not share my feelings. He wondered why such a thing had to happen just after he and Judy had found each other.

Jim and I continued to go to the meditation room after each visit. After praying for Judy and for strength to cope with her illness, Jim would take my hand and we would discuss her condition and share our thoughts about her.

During one of our discussions I said to Jim, "I know you and Judy had hoped to marry, but you really haven't known her very long." I didn't want Jim to feel he had to "stick by" Judy out of obligation.

"Time is irrelevant. When true love comes you know it right away," he said. "I think I loved Judy from the first time I saw her."

"But, Jim, do you know it might take a year or more for her to recover?"

"I've waited more than three months for her to come back

to me from Europe. I'll wait for her a year or even two if I have to."

I thanked God for bringing Judy and Jim together. I thought this was the proper time to give Jim what I had found on a page in Judy's journal. It said:

Stockholm, Sweden   October 13, 1975

To Jim,
  If my words could bring you across the miles and draw you with me in thought . . .
  They would speak of a love that reaches deep and of emotions within my being which are sensed by only you. As my tears fall I know that these tears go beyond fears and gladness and I dream of your tender touch, silent moments, heartbeats, quiet talks, laughter, and walks to nowhere.
  Suddenly, I'm with you, walking, talking, watching and sharing the beauty of this earth. Then, I come to the realization that we are apart in the dimension of distance but we are together in our thoughts of each other. Once again I know I love you.

Jim read it and I sensed it touched him deeply. He folded it and put it in his pocket.

We always felt a presence with us in the meditation room. On one occasion, after praying, we looked up and the lights were blinking softly. It seemed as if God were saying, "I am here with you. I hear you. I will take care of Judy." I felt a deep sense of security.

Gene told me that on one occasion, when he was feeling very low, the lights also blinked softly back at him and he had felt the same sense of security.

7

January 2   As Jim and I entered the room, Judy was staring vacantly. Jim stood beside her and said, "Hi, Judy. It's Jim. I can't tell you how much I love you and how much I've missed you since you've been gone. It's so good to have you back with me."

Suddenly Judy seemed to recognize Jim. Tears rolled down her cheeks. Jim, straining to keep from touching her, whispered huskily. "That's O.K., Judy. You don't have to cry. I'm right here with you."

Judy remembers:

Finally, after all these months, I was with Jim again. I had been away from him so long. But why couldn't I touch him, hold him? I cried.

Jim and I were ecstatic. We felt Judy must have recognized him. We rushed home to tell Gene, who was thrilled.

As Jim was leaving, he grabbed me and gave me a big hug and shook hands with Gene to express his joy.

Our doctor had told us we would experience terrific highs and deep lows during Judy's illness. That night definitely was one of our highs.

January 3   Jim came down to spend the weekend with us so he could go to the hospital and see Judy more often. When we were watching Judy she seemed to try to move her hands. Jim told her to move her hand up toward him and we thought she did, although she showed no sign of recognition.

January 4   We all went to church. Jim and I went to the hospital. The doctor was just coming out of Judy's room. He said, "I'm very concerned about Judy. She hasn't shown any sign of coming out of her coma."

Jim said, "She seemed to recognize me and to move her hand when I told her to."

The doctor said, "Those are just gross movements that mean nothing. Even though her eyes are open, she can't see anything. She's just staring into space."

Jim said, "I'm very objective when I watch her. I first watch her movements before we go into her room, and then I watch them closely when we're with her."

The doctor said, "When do you see her—once a week?"

Jim answered, "I've been here almost every day this week."

The doctor asked me, "How is the fund coming?"

I said, "I have no idea, but I'm much more concerned about Judy right now than about the fund."

"We don't know how much oxygen she lost to the brain," the doctor said. "She was in the room with the heater for so long. The longer she remains in a coma, the less chance there is of her coming out of it."

"But I have faith she'll come out of it," I said.

He said, "If she doesn't come out of her coma in two weeks, I'm going to have to write her off."

I wanted to shout at the doctor to stop talking and leave. I was so glad someone was with me. Jim and I both felt like we had been kicked in the stomach.

Jim said, "I want to go to the meditation room before we see Judy."

"So do I," I said desperately. "Let's go right now."

After a short time of prayer, we went back to Judy's room. Jim tried to talk to her as enthusiastically as he always did, but he couldn't. I couldn't seem to say anything to her either. Our spirit seemed to be gone, as if we'd lost all faith. Jim said, "I think we'd better go back to your house. We're not doing Judy any good here."

I didn't want to tell Gene what the doctor had said. He and Jerry both have high blood pressure and easily become emotionally upset. But he wanted to know why we came home so early, so I told him some of the doctor's words. He replied, "Well, I'm going to go to the hospital and see how she is for myself." He went alone.

Jerry had been laid off from work a few days before Christmas and had nothing to do all day but think of Judy. He seemed more quiet and withdrawn. I soon learned why.

After Gene went to the hospital, Jim told Jerry what the doctor had told us. Jerry said our doctor had called him two days before about a minor ailment of his and then had told him essentially the same thing. Jerry was crushed by the doctor's prediction. He was sure Judy would never come out of her coma or be normal again.

Each time Jerry had gone to see Judy, she'd had bad body tremors and had remained unresponsive. He became depressed and stopped going to the hospital. We told Judy that Jerry had a bad cold and was not able to come see her.

Jim stayed until mid-afternoon. Then he said, "I'm going back to Madison. I just can't go back to see Judy. I know I wouldn't do her any good." He was very dejected.

It was my father's birthday, which I had forgotten. My mother called and asked Gene and me to come to their house for coffee and cake. I was so depressed I didn't want to go, but I told her I would let her know later.

Gene came home pleased that Judy seemed more alert. He was very enthusiastic, but nothing he said could bolster my defeated spirit.

We did go to see my parents for a short time, but I wasn't very good company. We discussed what the doctor had said. My brother Harold's wife, Jeanne, said, "Why don't you tell some of the staff what you think you observe with Judy?" I decided to do it the next time I noticed anything different.

Gene persuaded me to go to the hospital in the evening. Judy looked more alert when we saw her. She seemed to be looking right at me. As we were leaving I told a nurse I thought Judy was really looking at me. She smiled with pity in her eyes. I could see she thought I was imagining things. I became depressed again.

I retired for the evening, feeling very low. Gene prayed with me, as he did every night. I was still feeling blue, when suddenly I seemed to hear a deep, sad voice say, "Oh ye men of little faith!"

I was ashamed for my lack of faith all day. I had been aware of all the small miracles since Judy's accident, yet I had doubted God would continue to take care of her. I asked God's forgiveness and never again doubted that she would recover.

8

January 5   I knew Jim would be feeling low so I called him and told him Judy seemed much better. He was relieved and said he would come see her later in the week. At a nurse's suggestion, we brought a radio to play in Judy's room all day, hoping the music would stimulate her brain.

We were pleasantly surprised when we received a Christmas card from Dr. Luis. He wrote, "Dear family Steuck, I arrived very well to Madrid. We want, I and my wife, very Happy Christmas and prosperous year 1976. Thank you very very much for all. Luis."

January 6   A former neighbor suggested that showing pictures to a comatose patient might help. We were grasping for straws, trying to stimulate Judy's brain as much as possible. I said, "Thanks for the information. I'll try to find pictures that will mean the most to her."

I thought of the family picture taken just before Judy left for Europe. I hoped it might stir up memories. She and Janet had laughed and kidded around as we were having it taken.

I had one large picture in color of the family and another of Judy and Janet. I chose the larger pictures because her pupils were always dilated and I knew she couldn't see well, if at all.

January 7   I held the family picture in front of Judy's staring, dilated eyes. Tears came rolling down her cheeks!

"Thank God," I thought. "She must be able to see, and she must have a memory."

Judy remembers:

I didn't know what had happened to me or where I was. I thought I was in heaven having flashbacks of my life with my family. The faces were fuzzy. It was like a movie. I couldn't get everything straight in my mind. I felt very sad because I wanted to be with my family again.

January 8   I called Carol Gevaart, the teacher with whom Judy had done her student teaching as a speech clinician, to see if she might have any suggestions for helping her. She said, "We're learning new things about the brain all the time. It's still a mystery."

Then she said thoughtfully, "When Judy tries to speak and realizes she can't, she's probably going to be very frightened. As a speech clinician she knows all the causes for loss of speech, and she'll probably wonder what could have happened to her."

I had thought the same thing and had been very concerned about it.

She paused and wondered aloud, "Perhaps if the children Judy taught were to send cards to her and add a personal message, you might see some reaction from her as you read the names—especially Kevin's."

I said, "Yes, I think that sounds like a good idea, and I'm sure she'll remember Kevin if she's able to remember anyone."

January 9   Judy had to be turned onto her stomach in her Stryker frame twice a day, once in the morning and once in the afternoon, to help her blood circulation. The nurses would place a canvas frame on top of her and deftly flip her 180 degrees. She was horrified each time. Her eyes got as big as saucers.

Through the opening for her face, Judy had only the floor to stare at. The nurses increased the time she was on her stomach from half an hour to two hours after she could tolerate it for a longer period. Judy disliked every minute of it.

I mentioned to the nurses that Jim didn't like to visit Judy when she was on her stomach. Whenever they saw me during the day they would ask, "Is Jim coming down to see Judy today?"

If I said yes, they would want to know the time. Then they would say, "We'll try to have her on her back when Jim comes. It's hard for him to talk to her when she's on her stomach."

When Gene and I arrived at the hospital, Judy was on her stomach. I was talking to her and rubbing her back when suddenly a horrified look appeared on her face. She seemed to be trying in vain to talk.

I said, "That's O.K., honey. You don't have to be frightened because you can't talk. You have something in your throat which keeps you from talking. You got sick while you were in Spain and I went over there and brought you home. You are with Dad and me now."

46

She didn't seem to understand. She still looked horrified.

I disliked being on my stomach. I wanted to say something about it. I tried to speak but I couldn't. I couldn't understand what had happened to me and why I couldn't speak or move. I couldn't comprehend what my mother was saying to me. I didn't know where I was.

January 10   In Judy's journal I found a wedding picture of her friend Kathy, whom she had visited in Austria. It was larger than a snapshot and I hoped Judy would be able to see it. I hoped that she would wonder why I had it and that maybe she would realize she was back home.

I put the picture up close to her dilated eyes. She seemed upset. There were tears in her eyes.

I thought, "That's my picture. How did it get out of my journal?" I tried to reach for it and I thought my arm went out, but I couldn't get the picture. I thought, "Where am I? What happened to me? I must not be in heaven after all."

January 11   The cards from Judy's students arrived. We read her each card with its personal message.

Kevin's personal message read: "Dear Judy, I hope you git better. I feel bad about it and right now I don't no what to say but git better now please and come and see me. Will I have to go now. I wish I could right more. Just git out of bed for me. Love you All ways. Kevin."

I remembered teaching all the children and how much I enjoyed it. I wanted so much to see Kevin and talk to him again. I wanted to get out of bed so badly. I wondered if I would ever be able to see any of the children again. I wondered where I was.

9

Gene's work hours were changed to 10:00 A.M. through 7:00 P.M., so he didn't have much time to see Judy in the evening. But he always went to the hospital before work and again during his lunch hour. He said, "I just have to see Judy in the morning so I know how she is. Even though she may not be as good on some days, I have to see her before I can start my day."

I had gone back to work after the first of the year. Gene always called me at work and told me how Judy was after he went to see her.

I would leave work about 3:00 P.M. and pick Janet up at school when she wasn't practicing for Pom-Pon Girls. We would stay at the hospital until 5:00 P.M. Then I would go back to the hospital in the evening with Jim when he was able to come. On Sunday Gene, Jerry, and Janet would go to see Judy before church. Jim and I went in the afternoon and evening. We stuck to our schedules so there was usually someone with Judy morning, noon, and night.

January 13   Gene and I were driving to the hospital for our evening visit. All of a sudden he started to cry. I wasn't alarmed, because he cried quite often. I said, "You're just going to have to get hold of yourself and not let this get you down."

But he had good reason to cry this time. Sobbing, he said, "I saw the doctor at the hospital this morning. He told me we had better plan to build a room onto our house to put Judy in."

I was shocked. "Why didn't you tell me when you called me at work?" I asked.

He tearfully replied, "I didn't want to worry you."

I felt sorry for Gene. I could see how defeated he felt by what the doctor had told him. I knew that feeling only too well, but he had been alone. All day long he wondered if all the little signs of recognition we had observed with Judy existed merely in our imagination.

"The doctor also said if the hospital is too full they'll probably have to put Judy out of ICU. Maybe we should think of a nursing home for her."

No way would I consider putting Judy in a nursing home —yet. It may have been some protective mother instinct, but I had faith that Judy would recover, even though it might take years of rehabilitation. I said vehemently, "She's not going to be put into a nursing home yet! If that's the way the doctor feels about Judy, I'm going to see about having her transferred to Madison."

Gene said, "The doctor also said he wants us to talk to the neurologist when he comes down in two weeks. They're going to do another EEG on Judy the day before he comes so he can discuss her with us."

I knew they would try to tell us Judy would never recover.

I decided to call a friend whose son was a medical student at the University Hospital in Madison to see if they could help us contact a neurologist there.

10

January 14   I called my friend, Charlotte Zimmerman. She and her family had moved from Beloit when our children were small, but we had remained good friends. Her son Don was a medical student at the University Hospital in Madison. He had donated a kidney to his mother.

Charlotte was deeply moved after I told her about Judy. She agreed to ask Don to talk to the head neurologist and get his opinion about transferring Judy to Madison.

January 15   Charlotte called back. Don talked to Dr. Manusher Javid, head of the neurological department at the University Hospital in Madison. He said that if Judy was being fed through her nose and had no bed sores, the best thing for her would be tender loving nursing care and the love of her family.

Judy was receiving constant tender loving care night and

day from the nurses and staff. Still in a Stryker frame, she was turned onto her stomach twice a day, and she had no bed sores. She was receiving 1100 calories of life-sustaining liquid each day through a tube in her nose. She was also receiving daily physical therapy to prevent muscle atrophy. We decided it was best for her to stay where she was.

But the neurologist also suggested that as soon as Judy came out of her coma she should be transferred to Madison.

Charlotte had been near death herself before her transplant. She said, "It is very important at a time like this to be touched and caressed and to know you are loved. Be sure you tell Judy how much you need her. Also, there are just certain people you really believe when they tell you that you're going to get better and it makes you feel more positive in your own mind."

We couldn't touch Judy unless we had gloves on, since she was in isolation, but we had been encouraging her and telling her that we loved her. I prayed that she would believe Jim and our family when we told her she would get better.

During this time Jim was poring over all the medical books he could find to learn about the brain. He learned that when a part of the brain has been permanently damaged (shown by a flat area on an EEG), another part of the brain can be taught to take over for the damaged part. This discovery was encouraging. We had faith that Judy would come out of her coma, but we knew there was evidence of brain damage. We expected to have to help teach her to talk, read, write, and walk all over again.

January 16   I told Jim that Charlotte had said that when she was ill it was important to her to know she was loved. That night at the hospital he said, "Judy, do you remember the question you asked me in your last letter?

There wasn't time to answer your letter, but the answer is yes."

I had no idea what Jim meant and didn't ask because I thought it must be something personal. Every time he saw Judy after that he always told her the same thing.

I felt I had been away from Jim for so long when I was in Europe, so in my last letter to him I asked him if he still wanted to marry me. As I became more aware of things, it made me very happy to hear him say yes. That knowledge made me eager to get better as quickly as possible.

January 17   Judy's eyes seemed more bright and alert. Whenever Gene visited Judy he would say, "Judy, how would you like to go to Florida when you get better? You'll be able to lie in the warm sun and get a nice tan. Doesn't that sound like fun?"

I had gone to Florida for the last few years in the spring and just loved it. I was so eager to lie in the warm sun and get a tan again. It all sounded so inviting.

January 18   During the night I had dreamed that Judy had recovered and was talking and laughing as she had before her coma. It gave me a good feeling when I woke up. When I mentioned the dream to my family, they all said they'd had the same kind of dream the night before!

In Judy's room that evening, I heard Jim tell Judy that he had dreamed about her the night before. I was shocked. When I asked him about it later, I learned that his dream was the same as ours. None of us had ever dreamed about Judy before. Was this a sign?

January 19   Judy continually kept her hands up tight against her chest. The nurse told us to try to keep her arms down as much as possible. As she slept, we would pull her arms down. As soon as she opened her eyes she would pull her arms back with tremendous strength and look angry.

I wanted my arms up to my chest because it seemed to be the most comfortable and natural place for them. I didn't want anyone trying to pull them down. It made me very angry.

Judy's pupils were less dilated than they had been. Her wrists moved more freely. She seemed more relaxed and aware of things. We continued to explain what had happened to her and where she was, hoping it would start getting through to her.

The children Judy had taught sent more cards. When we read them to her, we thought she gave greater response, especially when we read Kevin's message: "Judy, I wish you were here So I can tell you what I got for a grad in Science after you helped me. Well thats all. Please git better so you can come and see me. I don't no what to say so git better and come and talk to me. All my Love Kevin."

I really enjoyed getting all the letters from the children, especially Kevin. How I wanted to see Kevin and talk to him and be able to help him again! I was beginning to realize I was probably home.

January 20   I took an enlarged picture of Judy and Jim to the hospital. When she saw it Judy closed her eyes tightly and big tears rolled down her cheeks.

I was so happy having a picture of Jim and me. The picture had been taken at a wedding reception where I had been a participant and it had been such a happy time. I wondered if we would ever be able to have any more good times together. My mother had taken the picture shortly before I left for Europe and I had never seen it before.

Nurses started to close off Judy's tracheotomy at 15-minute intervals. A nurse was always present when it was closed, in case Judy choked, but she seemed to tolerate it well. We tried to get her to talk, with no results.

January 21   Gene always pinched Judy's leg when he came to see her. He wanted to see if she would react. Her leg would move very slightly, indicating she must have feeling in it. It always seemed to assure him. Today she moved her leg herself. We were very pleased.

I put my hand in Judy's and said, "Judy, can you squeeze my hand?" I felt a very slight movement.

"That's just great," I said. "You squeezed Mom's hand very well."

Gene tried the same thing and he was sure he felt some movement too. Tears filled his eyes.

I was finally beginning to realize my plight, but I still couldn't understand what had happened to me. I tried so hard to squeeze Mother's and Dad's hands, as they asked me to do. I thought I did, very hard. I also wanted to tell Dad to please stop pinching my leg all the time.

At times Judy would stare out of her window in a deep trance. We were unable to get her out of it unless we stood in front of the window to block her view. She would then look startled, as if she wondered where we had come from.

I was looking for familiar surroundings to see if I could figure out where I was. I thought I must be back home.

January 22   Harold visited Judy for the first time since she came home. Gene and I had told him about the hopeful signs we observed in her. He wanted to find out for himself if it was just our imagination, as everyone thought.

Harold told us, "No one was present when I visited Judy. I talked to her and held her hand. I noticed a slight indication that she might be trying to squeeze my hand. I wondered if there could possibly be any hope for her. It didn't seem possible, but I prayed for hope."

I remember Harold coming up to see me. I was glad to see him and I didn't want him to leave me because I was lonely. I tried very hard to squeeze his hand as he asked.

Later he told us, "A friend of mine, a doctor at Beloit Memorial Hospital, told me that there is absolutely no hope whatsoever for Judy. Her doctor consulted with him and some other doctors and they agreed that all medical evidence indicates she'll be nothing but a vegetable the rest of her life."

But there *was* another way. The doctors had forgotten someone more powerful than they who is able to perform miracles. Each Sunday our church had special prayers for Judy, and people all over the country and in Spain were praying for her recovery.

Janet had suffered her own inner turmoil for a long time without telling us. She was a sophomore in high school, and when she had returned from her Christmas vacation a boy had said to her, "What's the real story about your sister? I hear she really O.D.'d."

Janet angrily replied, "Don't you read the papers?"

"Nah," he answered curtly, "but I hear she's going to be nothing but a vegetable the rest of her life, anyway."

In frustration Janet hit him and shouted, "She is not!"

There were other similar conversations with unthinking young people. Janet kept her sorrow to herself for a long time because she didn't want to burden us with anything more. I was unaware of her different mood because we were so busy going to the hospital trying to help Judy. When she finally told me about it, I felt compassion for her because she had carried the burden by herself for weeks.

January 23   Judy never had any facial expression. We had to judge all her reactions from her very expressive eyes.

Jim decided to find out if Judy could read. He wrote on a piece of paper, "I love you more today than yesterday." When he held the paper in front of her face, her eyes lit up and seemed to smile.

I was so happy to see what Jim had written. Although he had been telling me every time he came to see me that he still loved me, it made me feel so good to see it in writing—especially those words, because they were from a card I had given him. I thought I gave him a great big smile.

January 24   I had been told that I should start reading children's books to Judy, since she would probably have the mind of a small child when she came out of her coma. I had kept some books our children had when they were small and was going to get them, but someone held me back. I could not start treating her like a child instead of an adult.

Judy later showed me her book titled *Diagnosis and Evaluation in Speech Pathology,* which states:

> Children's books are insulting to aphasic (brain-damaged) patients. Even though severely retarded in language usage they retain an adult outlook. It is essential to deal on an adult level. It is infantilizing enough to be physically helpless and to have lost the power of speech.

Thank God I did not read Judy children's books. Instead I found some poems and readings in Judy's journal I thought would be meaningful to her. One I especially liked and I read it to her often:

> There is nothing more beautiful than a rainbow. But it takes both rain and sunshine to make a rainbow. If life is to be rounded and many colored, like the rainbow, both joy and sorrow must come to it. Those

who have never known anything but prosperity and pleasure become hard and shallow but those whose prosperity has been mixed with adversity become kind and gracious. —Author unknown

That was a favorite reading of mine and really meant a great deal to me. My mother kept reading it to me. As I drifted in and out of awareness I would see a big beautiful rainbow and I would wonder at first if it had rained.

January 25   Judy seemed quite alert in the morning, but when Jim came to see her after driving from Madison through a blinding snowstorm, she seemed very tired and unresponsive.

Judy had graduated from the University with a B.S. degree in speech pathology and was a licensed speech clinician. When I noticed that she seemed to be yawning a great deal, I asked another speech clinician if he thought it meant anything. I thought Judy might be aware of what she should do to start speaking, because she was a very good student and enjoyed her studies. He said that it probably just meant that she was tired.

I finally realized I couldn't talk because of my tracheotomy. I remembered from what I had studied so earnestly that yawning would strengthen the tongue muscles, so I started yawning often.

January 26   Judy seemed very depressed and started to cry out loud.

I knew I was very sick and became depressed because I knew it would take a long time for me to recover. It was also very depressing to have so many people treat me like a baby instead of an adult and at times talk about me as if I weren't there. I wanted to tell them I wasn't a baby and I heard what they were saying. I remembered that crying would help strengthen my vocal cords so I would cry, sometimes very hard. I felt like I was a prisoner of my own body, unable to speak or move.

January 27   Our doctor arranged for Gene and me to talk to the Madison neurologist. When we arrived our doctor was with Judy. I had not seen or talked to him since January 4, so I said, "I've been keeping a log of Judy's progress."

He answered, "You'd better tell the neurologist. I have to go back to my office."

When we talked to the neurologist I told him about my log. He said little and offered little encouragement for Judy.

Judy's body seemed tight and she was very depressed. The neurologist recommended a new muscle relaxant.

The neurologist told our doctor Judy could have a skin graft on her stomach. The doctors had been concerned about her tolerance of the anesthetic. The surgery was scheduled for February 2.

11

January 28   The woman in charge of special services at the hospital called me and said she would like to talk to me about Judy's insurance and her mounting expenses.

When I arrived the administrator of the hospital was also in the office. He expressed his sympathy and concern for Judy and said he also had children about Judy's age. He said, "I know just how you must feel." But then he stopped short, and thoughtfully said, "No, I really don't have any idea how you feel. I couldn't possibly know."

I had bought medical insurance for Judy through the university before she left for Europe. The coverage was for $30,000, which at the time sounded more than adequate for any emergency. But Judy had already been in ICU for more than a month and expenses were mounting.

The woman pointed out that Judy would probably be hospitalized for many more months and $30,000 would not go very far these days. She said, "Your hospital bill is already $14,000 and Lord only knows what your doctor bill will be." She mentioned several agencies that might be of help.

Then she said, "I feel very remiss because we have not done more to help you and your family, but we just don't know what to do."

I thanked her and went right up to see Judy, but I was depressed and did not stay long.

I had found the best way to cope with a situation such as ours was to put complete trust in God and just live one day at a time, taking whatever each day brought but trying not to look very far into the future. But now I was forced to look ahead and I wondered how we were ever going to handle the mounting expenses.

When Jim and I went to the hospital that evening I told him the administrator had said he thought he knew how we must feel but had changed his mind.

Jim said, "That's something none of my friends have ever said to me. In fact, they all say they have no idea what it must be like, talking for hours to someone who can't respond. They also say they can't figure out how I can be sure Judy knows me. I tell them I don't know how I know, but I do know she recognizes me and is aware I'm there."

I had observed Judy's response when Jim came. I said, "Jim, when you come into the room Judy's breathing is different. She always breathes harder when you're there, and the look in her eyes is different from the look she has when Gene and I are there. It's the look a woman gives only to the man she loves."

Jim and I went into Judy's room, and Jim noticed how she breathed hard when he arrived. She seemed very happy to see him. Jim thought she was the best he had seen her since her accident.

When Jim was talking personally to Judy I always stayed

at the foot of the Stryker frame so it would seem more like they were alone. Whenever Jim ran out of things to tell Judy he would say, "Judy, your mother wants to talk to you for a while." Then I would get all the mail for the day and start reading all her cards to her. If she had not received much mail I would read cards I thought would be more meaningful to her over and over again, hoping one of the messages would reach her.

Jim had said, "Sometimes I feel as though I'm repeating myself so much, but I don't know what else to say sometimes."

I answered, "I wouldn't worry about that, Jim. We don't know what is getting through to her and what isn't, so it can't do any harm."

My mother read the same cards over and over to me and Jim kept saying the same things to me day after day. I wanted to say, "I've heard all that before." I felt as though I could recite everything they had said word for word.

Again that evening I told God I was going to place all our worries in his hands. I went back to living each day as it came, knowing God would help us take care of our future needs, whatever they might be.

January 29   Judy has pretty white teeth, although the cap on her tooth had been knocked off in Spain. She had always brushed them often when she was well. Since her accident the nurses had been swabbing her teeth with a sponge. One day a nurse suggested we get Judy a baby toothbrush and she would try brushing Judy's teeth.

Gene brought a toothbrush to the hospital. The nurse brushed Judy's teeth, then put a little water into her mouth. Suddenly Judy started to swish the small amount of water around in her mouth. Gene was amazed.

It seemed so good to get my teeth brushed again. I knew

*it had not been done for a long time. It was just natural for
me to swish the water around in my mouth.*

Jerry hadn't seen Judy for three weeks. I finally talked
him into going. I wanted him to see for himself that she
really was responding to us. He said if he left suddenly I
was not to be alarmed, because it would be because he could
not bear seeing her in her condition.

When we arrived in Judy's room I said, "Judy, why don't
you move your leg so Jerry can see you do it?" She moved
it very slightly.

"Isn't that great, Jerry?" I asked. I looked up. Big tears
were rolling down Jerry's cheeks. He finally realized she
was responding to us after all. From then on he went to the
hospital as often as he could.

*It seemed so good to see Jerry again. I wanted to ask
him about his cold. My parents had said that was what kept
him away for so long.*

January 31   Most people were convinced that what we
saw in Judy's eyes and in her movements meant nothing,
that it was just wishful thinking.

*I did not respond to anyone but Jim and my family be-
cause hardly anyone else treated me like I could understand. I could tell Jim and my family believed I would get
better.*

Jerry kept saying to Judy, "We all know you can hear us
and we know you can understand us. Don't pay any atten-
tion to anyone who treats you like a baby and thinks you
don't understand."

Jerry bought a cassette player for Judy. He taped songs
he thought she would like and played them for her. She
seemed to enjoy them. We asked the nurses to play them
for her when no one was with her.

Judy moved her legs a little more. Jerry was thrilled with

every little thing she did. She seemed to try very hard to say hi.

I enjoyed hearing all the songs Jerry had taped for me and having him come to see me so much more. I even tried to hum the songs. I tried to say hi but nothing came out.

February 1   Judy always followed people with her eyes, although she never moved her head. When she was put on her stomach in the afternoon she cried very hard. The nurse on duty felt sorry for her so she rolled the Stryker frame to the doorway and Judy watched the activity in ICU.

I disliked being on my stomach so much. I also dreaded being left alone. I found that whenever I cried the nurses would come and try to comfort me and would stay with me for a while. It was also another good excuse for me to cry so I could strengthen my vocal chords.

When the nurse moved me to the door of my room I thought I was in the main lobby of the hospital. I saw all the activity and I saw flowers sitting around. I thought I saw the gift shop and I looked all around to see if I could see the snack bar because I had always heard they made delicious shakes.

Jim loves to hunt and go ice fishing in the winter, but because he came to the hospital so often he was unable to go ice fishing very often. I persuaded him to go Sunday with his friends because he was going to be at the hospital the next day after Judy's skin graft. It was the first Sunday afternoon and evening Jim and I had not stayed at the hospital with Judy. Janet and Jerry went in our place in the afternoon.

Gene and I went to the hospital in the evening and I told Judy Jim had gone fishing and would not be able to come to the hospital to see her until the next day. She started crying so loud the nurse came to see what had happened. She would not look at me for the rest of the evening. She would respond only to Gene. I felt very hurt.

I always looked forward so much to Jim coming to see me. I was disappointed he had gone fishing instead of coming to the hospital as he usually did. I was angry with my mother for letting him go. It was also another very good excuse for me to really cry. I didn't care who heard me.

February 2   I called Jim in the morning and told him how Judy had reacted the night before. He felt bad and said, "You probably never should have told her I went fishing. You should have said I just had to go out of town. I didn't think she would get that upset."

I replied, "I didn't expect her to react that way either."

"I should be able to get to the hospital about 3:30 today, so I'll see you then," Jim said.

The skin graft was done on Judy's stomach in the morning. She came out of the general anesthetic well. Skin for the graft was taken from her upper right thigh, front and side. A culture was taken of her stomach after the skin graft. No foreign germs were found. At last she could be taken out of isolation! The nurses put all objects in her room, including the scissors, into a large plastic bag to be incinerated.

We no longer had to wear robes or gloves in Judy's room. It seemed like a real turning point. For the first time we saw real smiles on the faces of the doctor and nurses.

Pastor Nycklemoe was in Judy's room when I said, "Judy, Jim will be here to see you very soon now." He was amazed at the look of love that filled Judy's eyes.

Jim was delighted that he didn't have to wear a robe. Judy was very glad to see Jim, but she seemed to be in a great deal of pain. Jim stayed with her for several hours.

The physical therapist had been concerned about Judy's reaction when he worked with her left leg. She seemed to be in pain when he lifted it. He wondered if she could have hurt it in Europe. We'd heard nothing about it in her letters.

We finally received Judy's luggage. It had remained in Amsterdam for a month and then was sent to California by mistake. We put it upstairs with her other packages and luggage.

February 3   Gene went to the hospital in the morning. He said Judy's eyes seemed bright, but he was sure she did not want to be left alone. I called Jerry and asked him if he would go to the hospital at 1:00 P.M. and stay until I could get there at 3:30. He said, "I'd be glad to, Mom."

When Jerry was with Judy, Dr. Twyman, the otolaryngologist who performed the tracheotomy, came to see her. Her trach had been removed a few days before. When he examined her throat, she started to cry very hard. He was concerned about her loud crying and said he was going to give her a hearing test. He was afraid that she was not only affected mentally but had also suffered loss of hearing. We were confident she could hear but thought she was afraid of the doctor because he had hurt her when he removed the trach.

When Dr. Twyman came in to check my throat and acted as if he didn't know me and talked about me as if I weren't there, it was more than I could take. He was the only one of all the doctors or nurses I knew personally. I was a good friend of his daughter. It didn't bother me quite as much when people I didn't know treated me that way, but when someone I knew acted as though I weren't even there, I cried from frustration.

Jim came in the evening. Judy seemed very tired, but she never cried when Jim was there and she always seemed happy to see him. She watched everything he did with eyes filled with love.

Jim bent down and kissed Judy twice on the lips. As always, her mouth remained open and there was no response. Then Jim kissed her hand and said, "Judy, I want you to kiss my hand as I kissed yours."

He put his hand to her mouth and all of a sudden she responded—she kissed his hand!

Jim bent and kissed her on the lips, and she closed her mouth and kissed him back. Jim couldn't stop kissing her. Tears filled his eyes. He said, "Judy, you're making me cry, I'm so happy."

Finally Jim kissed me. It had been so long. I thought I had responded to his first kiss. Then I really did kiss back. It felt wonderful to really be with Jim again and have him kiss me.

February 4   Judy was very bright-eyed and she tried to talk. She continued to cough, yawn, and cry out loud. She cried so hard people would stare as they went past her room.

Judy's arms seemed more mobile and she seemed more responsive to the physical therapist. She did as she was told, although she seemed to be in pain.

When Dr. Twyman came to see her she cried again. She always cried when she saw him. He felt very bad about her reaction.

Jim came in the evening and Judy responded to his first kiss and seemed very emotional.

We wanted to see if Judy could read. I put her glasses on

her and Jim showed her the Christmas card he had bought for her but had been unable to give her.

Jim said, "Judy, I'm going to show you this card. I want you to read the front of it and when you're finished I want you to blink twice and I'll show you the inside."

We watched as her eyes moved across the words, "For my sweetheart at Christmas." Her eyes moved down to the verse. Then she blinked twice.

Jim eagerly opened the card and we watched her eyes move across the words, "Merry Christmas, with all my love. Only you can fill my heart, my needs, my world."

Judy looked very happy. Jim and I were convinced she had read the card.

Jim had gotten me such a beautiful card. It had such a meaningful verse. I felt so happy. I wanted so much to be able to talk so I could tell him I loved him.

Jim said, "Judy, would you like a drink of water? If you do, blink twice and we'll get the nurse and ask her to give you one."

Judy looked directly at Jim and blinked twice. I went and found the nurse and said, "Judy would like a drink of water."

The nurse looked very puzzled and asked, "How do you know that?"

"We told her to blink twice if she wanted a drink," I answered, "and she did."

The nurse came right in with a rubber syringe. She put some water in it and put it to Judy's mouth, but Judy wouldn't let her put it into her mouth. I'm sure the nurse thought we were imagining things.

I didn't want a drink from a rubber syringe, so I wouldn't take it when the nurse put it to my mouth. I wanted a regular drink of water.

After the nurse left, Jim said, "Judy doesn't want a drink

from a rubber syringe. Who would want a drink from that?"

Jim put a few drops of water into a pill cup and put it to Judy's mouth. She opened her mouth and drank it.

I enjoyed my little drink from the tiny cup, but it was hardly enough to quench my thirst.

12

February 5   Jerry was reading through some of Judy's speech therapy books and came across one that had a great deal of information about learning to speak again. He told Judy he had read one of her books. Then he read portions of what he had copied from the chapter "Voice Therapy for Function Aphonia":

1. Explain the problem to patient and discuss it.
2. Do not imply to patient she could phonate normally if she wanted to.
3. Evaluate whatever nonverbal phonations the patient may have in her *coughing* and *crying* and describe them as normal phonatory activities. (Where the vocal cords are getting together well to produce these sounds.)
4. These nonverbal phonations should then gradually be shaped into use for speech attempts to nonsense syllables, and then move on to single words, but with no early attempts at phonating during real communication.

The book said to tell the patient to cough and yawn. "If the vocal cords are apart too long we sometimes experience trouble getting them together to make a voice, although we still can make other sounds.

"First a/ and o/—open mouth vowels b, d, i, y—all, are, oh, oak, open, arm—cords are getting stronger."

I was so glad to hear what Jerry read to me. It was from one of my favorite speech therapy books, The Voice and Voice Therapy *by Daniel R. Boone. (I could never forget the author's name.) I had gotten so much information from it regarding voice therapy. When Jerry read from it doors opened in my mind. The book also had a lot of information about the tongue and how important yawning is in strengthening the tongue muscles.*

Jerry started his first lesson as a speech clinician. He said, "Now Judy, I want you to make the a/ sound," and he showed Judy with his mouth how to make the sound.

Judy immediately responded with "a/."

Jerry was so excited he slapped his hand on his leg and said, "Now can you make the o/ sound, Judy?" He put his face by hers and showed her how to make the o/ sound.

Again she responded immediately, this time with the o/ sound.

Jerry was elated. He didn't know he would be such a good speech clinician.

After I watched Jerry's mouth, as he showed me how to make the a/ and o/ sounds, I shaped my mouth as he did and made the sounds also. He was so intent on his work. I knew how unfamiliar it was to him, and I laughed to myself, because I knew exactly what I should be doing now.

Jerry then said, "I suppose you're wondering why none of your friends have come up to see you."

Judy completely surprised Jerry by answering "Y?" (another vowel sound).

Jerry told her she was in intensive care and no one but family members were allowed to visit her.

I had wondered why my friends had often sent cards, yet had never been to the hospital to see me.

Jerry asked Judy if she would like to have him learn sign language so he could communicate with her that way for a while. Frantically, she blinked her eyes. He got the message. There would be no sign language.

I didn't want to speak in sign language. I wanted to work on getting my speech back again. I wanted to talk again as soon as I could. I was glad Jerry understood.

When I came to the hospital in the afternoon, Jerry happily told me of the progress he and Judy were making. He told Judy to make the vowel sounds for me, first the a/ sound and the o/ sound. She responded immediately. He showed me what he'd read from her book.

I was so thankful. It was just as I had prayed. I realized everything Judy had been doing—yawning, coughing, crying—was to enable her to speak again.

I said, "Judy, you're smarter than any of us. You knew exactly what to do to get your speech back and you did it." I gave her a big kiss and she kept blinking her eyes at me. She seemed so relieved we knew what she was trying to do.

At last! My family was aware of what I was trying to do to regain my speech. All I needed was a little help with it.

February 6   Gene made his usual morning trip to the hospital. As he was talking to Judy, she reached up weakly, touched his face, and said softly, "Hi." He was overwhelmed. He was still crying from happiness when he called me to share the news.

Jim went to see Judy at 3:30. He stayed with her until the PT (physical therapist) arrived. I stayed home and prepared our evening meal. When the PT came Jim left and came to our house for dinner.

Later the physical therapist told me that when he started to work on Judy's leg, she looked like she was in pain and she said, haltingly, "You . . . don't . . . think . . . I," but she stopped and was unable to continue.

I was going to say, "You don't think I can hear you but I can and I'm going to tell my dad you're hurting me."

When Jim arrived at our house for dinner, I said, "Did Judy say anything to you today?"

"She said three words," he answered.

Jerry asked, very excitedly, "Three words?" I could see he was trying to think what three words the book had said Judy would probably say next.

I said, "Jerry, I think the three words Judy said were 'I love you,' weren't they, Jim?"

With a big smile, Jim said, "Yes, that's what she said to me."

Jerry was confused. Judy didn't seem to be following his lesson plans.

We were all very happy with Judy's progress. I could hardly wait to return with Jim to the hospital in the evening.

When I had first started going to the hospital with Jim, I thought he might like some time alone with Judy so he could talk more personally to her, so I said, "Jim, if you ever want to be alone with Judy, just let me know and I'll be glad to wait in the waiting room for a while."

He answered, "I don't mind having you in the room with me when I'm talking to Judy. Actually, it gives me moral support. But if the time does come when I think it would be better to be alone with her, I'll let you know, O.K.?"

Jim and I went to the hospital in the evening and Judy

seemed to be trying very hard to say something. Jim decided now was the time for them to be alone. He said, "Judy, how would you like to have your mother leave for a while so we can talk by ourselves?"

Judy's eyes lit up. She seemed very happy with his suggestion.

I had become very aware of my mother's presence when Jim would kiss me and talk personally with me. It made me feel uncomfortable. I was so happy to have the chance to be alone with him.

I also felt this was a good time to leave Judy and Jim alone. I went to the waiting room, trying to read one magazine after another.

Later when Jim came to the waiting room, he said Judy had kept saying "Remember," but she couldn't say any more.

I was trying so hard to talk to Jim about some of the fun we'd had together, but nothing would come out.

13

February 7   One of the nurses called at 8:30 A.M. "Judy has started to talk," she said excitedly. We rushed to the hospital.

Judy's nurse, Barbara, had walked into her room, and Judy astonished her by simply saying, "Good morning."

All night I had practiced in my mind the tone of voice I would use to say "Good morning." I was eager to try to do what seemed to be a very normal thing. When Barb came into my room, I said, "Good morning." She looked surprised. She said, "Good morning, Judy." She left the room right away and came back with other nurses.

Barbara told us later, "It just about blew my mind. I ran out of Judy's room and got the other nurses. We all went into her room and told her our names and asked her to repeat them. She did, even Isabelle's."

Judy was tired by the time we arrived and she was not

talking or doing anything. The nurse said they had probably worn her out. Suddenly, a weak little voice said, "I'm thirsty."

I turned in amazement to Judy. It was the first time Judy had spoken directly to me for five and a half months. I said, in wonderment, "Would you like a drink of water, honey?"

"Yes, please." It was her first real drink of water since December 17. She didn't seem to want to stop.

At last, I could have a regular drink of water instead of just a few drops from a cup or from a rubber syringe. It tasted so good I didn't want to stop.

Then Judy looked at me and said, "Remember this day."

How could I ever forget February 7, 1976? It was the day God answered all the prayers for Judy and performed the "big miracle," releasing her from imprisonment in her body. I was so thankful.

We had another cause to rejoice—Judy's skin graft was a success. There were no signs of rejection. Oh happy day!

About two weeks before I had asked Judy if she wanted me to get the film from her semester in Europe developed. I thought she might like to see the pictures, but the look in her eyes told me she didn't want me to have them developed, so I never mentioned the film again.

I was very surprised when Judy said, very hesitantly, "Don't develop my film. I want to show my own pictures."

"Don't worry, Judy," I said. "All your film is still at home and you can get it developed yourself whenever you want."

Those were my pictures. I did not want anyone to see my pictures before I did. I was so worried my mother had gotten them developed because I didn't remember her saying whether she was going to or not.

I had called Harold to tell him our exciting news. He and Jeanne came immediately. We were all standing by Judy when she started to yawn. I explained, "Judy is yawning to strengthen her throat."

We were all stunned when a little voice corrected me: "No—the tongue."

We all looked in amazement at one another. Judy had known exactly what to do to get her speech back. I looked at Harold. Tears were rolling down his cheeks. This was the young woman our doctor and his doctor friend had said would be a vegetable for the rest of her life. It was truly a beautiful miracle!

Then Judy said, "I have to go to the bathroom."

I was really surprised. I rushed out to get the nurse. She hurried to Judy's room with the bedpan and asked us all to leave.

It didn't take Judy long to figure out that whenever she wanted someone to leave, all she had to do was ask for the bedpan. She usually did this when the therapist started working on her leg.

Later Judy looked at Janet and said, "What's happn'n, Janet?" Janet thought that was "pretty cool."

Then Judy said, "Is the mail here yet?" There were more looks of amazement. She had been aware of our reading mail to her daily! The mail hadn't come yet, but someone must have heard Judy ask, because it was brought in almost immediately.

I took the first card from its envelope. I showed her the front and read it to her, as usual. Then I opened the card and showed her the inside. I was going to read the message when I heard her say, "Dear Judy, We're anxious to see you."

Janet said in astonishment, "She's reading the card!"

It is very hard to explain my feelings at this point. The experience was beyond words. We all thought we would have to teach Judy to do everything all over again. But here she was talking clearly and coherently, and now she was reading.

"O my God, how wonderful thou art," I said.

I went home for a short time, although I didn't want to leave. I called Jerry and told him all the things Judy was saying.

Jerry was ecstatic. "I didn't know I would make such a good speech therapist! But she's not supposed to say all those words yet. The book said they don't come until much later."

I called Pastor Nycklemoe. He was thrilled to be part of something so wonderful.

After Pastor Nycklemoe went to see Judy, he said, "I felt absolute wonder, thanksgiving and awe. Tears came to my eyes as I walked into the intensive care unit and met the teary eyes of the nurses. Everyone in the hospital was crying for joy. When I walked into Judy's room, I said, 'Praise God, from whom all blessings flow.' I moved to the bed and with eyes open prayed a prayer of thanksgiving. Yes, a miracle of healing had happened right in our midst. Judy was going to be O.K. God had performed another miracle."

I also called Jim and told him our great news. He was very anxious to see Judy. Jim and I had made plans two weeks before to have his parents come down and see her. We were both happy that she would be out of her coma for their visit the next day.

Jerry came to the hospital. Judy said something we didn't understand, but Jerry thought it was off-color. He asked, with a puzzled look, "What did you say, Judy?"

Jerry's look struck Judy very funny. She started to laugh. It was the first time we had seen a smile on her face since September. Her laugh was the best sound ever. We all started laughing, almost hysterically. The nurse came in, and she started to laugh also. Judy's laughing was contagious. It was complete euphoria.

Dr. Twyman came in to see Judy. He said, "Hi, Judy. I'm Dr. Twyman, Lee's dad. I left my mark on you" (the tracheotomy).

Judy said, "Yes, I know. How are you, Dr. Twyman?"

It seemed so good to be able to talk to Dr. Twyman and have him say hi to me and act like he really knew me. Now I could talk with him about Lee.

I asked Dr. Twyman where Lee was and what she was doing. Before he could answer Judy told us where she was the last time she had heard about her and what she was doing. She also told us what his other daughter was doing.

Dr. Twyman was amazed to learn how sharp Judy's mind was. We talked for a short time and he said, "I'm going to have to get out of here." He was near tears.

When I went out into the hall with him, he said, "All Lee did on Christmas was cry about Judy. I asked her if she would like to see her for a few minutes, but she didn't think she should. And to think I was going to give Judy a hearing test. It must have been all the wonderful nursing care she received that did it."

The nurse standing by us said, "I'm afraid not, Dr. Twyman. It was the love of Jim and Judy's family and all the prayers that did it."

(We never received a bill from Dr. Twyman for any of his services. We knew how much he cared about Judy and have always been very grateful to him for all he did for her.)

I told the nurse that Jim's parents were coming the next day to visit Judy. I mentioned that Jim's father had never met her. She thought it might be a good idea to have Janet pluck Judy's bushy eyebrows. She said, "Judy, how would you like to have Janet pluck your eyebrows?"

Judy said, "That would be funny."

Janet laughed because she knew Judy never thought she was very good at plucking eyebrows. We just left them bushy.

I felt sorry for Gene because he had to be at work and wasn't able to witness all the wonderful things that took place that day, although I kept him posted by phone.

When he arrived at the hospital in the evening he tried hard to get Judy to say something. She cried, yawned, and coughed, trying very hard to strengthen her unused muscles so she could talk to Gene, but she was too exhausted to carry on a conversation. She said weakly, "I'm not strong enough."

Gene wanted so desperately to hear her talk, he kept trying to get her to say something. Finally in exasperation she started to stare out of the window as she did when she was in her coma. We realized how very tired she must be and left so she could get some much needed rest.

February 8   Gene, Janet, and Jerry went to see Judy before going to church. Gene finally heard Judy talk. He was very happy.

Gene mentioned something about a rose to Judy. She asked, "What is a rose?"

Gene later told me he was concerned because Judy couldn't remember what a rose was.

We all went to church and offered our thanks to God for the beautiful miracle he had performed.

When Pastor Nycklemoe greeted the congregation at the beginning of the service he simply said, "Praise God! A miracle has been performed before our very eyes. Yesterday Judy came out of her coma."

Everyone began to talk at the same time. The whole congregation was touched by the miracle. Tears of joy flowed from many eyes. The service was filled with joy and praise. We all knew a miracle had been performed.

When we mentioned to Pastor Hegre, our visitation pastor, that Judy did not know what a rose was, he suggested we take a few roses from the church bouquet and show them to her.

As soon as we entered Judy's room I gave her the roses. She didn't seem to be impressed. I thought she probably had forgotten she had asked what a rose was.

I knew what a rose was. I was just trying to remember Shakespeare's line, "A rose by any other name would smell as sweet." When my dad mentioned a rose it just brought the writing back to my memory.

Judy had never met Jim's father, so she was eager to look her best. She was not being fed any solid foods, so she still had a tube in her nose, but the nurses had washed her hair and brushed it and she looked the best she had looked for a long time, although she still had her bushy eyebrows.

Judy recognized Jim's mother immediately and told her she was happy to see her again. Then she was introduced to Jim's father. She said, "How do you do? It's nice to meet you." Jim's father was impressed with Judy's sharp mind.

Jim's mother brought Judy her first gift, a musical glass figurine of a little girl sitting on a bench next to a piano. Judy loved to hear it play. She said, "Oh, look how it goes around. That's nice."

In a short time Judy seemed quite tired so Jim's parents did not stay long. In the hall I said to Jim's mother, "Jim has been wonderful with Judy. He's been so patient with her and very encouraging."

She looked surprised. She said, "But Jim doesn't usually have any patience at all." I felt that God made sure Jim had patience when he really needed it so he could help Judy.

Happy with Judy's sudden recovery, Jim kept saying, "This is so unreal. It's so much like my dream. It's hard to believe."

"Isn't this an amazing miracle?" I said. "All I keep hearing in my mind is 'Amazing Grace.' "

After we returned to Judy's room, she looked at Jim and said, "Are you still saying yes?" (in answer to her letter).

Jim replied, "Yes, I am, Judy."

When Jim and I returned to our house we learned that Gene had done a taped interview with WBBM, Chicago, and it was to be aired soon. Everyone was sitting around the radio waiting to hear it.

There also was a message from UPI. The news of Judy's recovery was traveling fast. The whole weekend was filled with euphoria and jubilation.

14

The next morning the phone started ringing. Several radio stations requested taped interviews with us. It was very exciting.

I called my supervisor at McCleary Industries to say I would be late because of all the calls. Everyone at work had been kind and understanding to me during Judy's coma, especially when I would inadvertently make a mistake on one of the employee's checks when I was on the computer. I sometimes could not seem to concentrate on my work because of my concern for Judy.

When I arrived later everyone in the office was talking about Judy. She had worked there for six summers and was well known. My co-workers had lived through the ordeal with me and had offered much comfort and support.

About 11:30 Gene called and said he had talked to Channel 6, Milwaukee, and had agreed to an interview in our home at 12:30. I rushed home to straighten the house before the camera crew arrived. During the interview we thanked everyone for their love and concern for Judy.

Gene went back to work and the television men followed me to the hospital. They wanted to take some still pictures of the pictures we had shown Judy during her coma.

Judy was sleeping when I went into her room. I took the pictures down very quietly. I did not want to waken her because I had to go back to work.

NBC called in the afternoon and wanted to do a television interview with Gene, Jim, and me, but we couldn't all get together at the time they wanted so the idea was dropped.

When Gene had seen Judy in the morning, she was happy and talked a great deal with him. But in the afternoon Jerry found her depressed.

Jim was at the hospital with Jerry when I arrived. He met me in the hall and said, "Judy doesn't want to see you or me because she said all we do is talk about her."

I answered in disbelief, "But I've never said anything to Judy about talking to you about her. Why would she think a thing like that?"

He answered, "I don't know. Maybe I said something and didn't realize it."

Many times when Jim and I went to see Judy she had just been flipped from her stomach to her back. Being on her stomach seemed to tire her so much she would go into an exhausted sleep and we couldn't arouse her. We would then go to the cafeteria and talk, mostly about Judy. We always told Judy we were going to the cafeteria to have a cup of coffee and would be back after she had rested.

My mother had gone to the hospital to see Judy before she went back to Florida in the middle of January. She wanted Judy to know that Jim and I had become good friends, so she jokingly said, "Judy, you'd better watch out for Jim because your mother seems to like him very much."

She hoped that if Judy was able to hear, her comment would make her happy to know I thought a lot of Jim.

I was so upset by what Grandmother said. I knew my mother and Jim were always leaving to go to the cafeteria and I thought all they were doing was talking about me. I was very resentful of the time they spent together away from me. I felt I should have been the one to be with Jim, talking and having coffee with him, not my mother. I became very jealous of her for the time she was spending with Jim.

When Jim and I went back in the evening to see Judy, she was happy to see us and we had a nice visit. We could not understand why she had acted as she had.

Later Jim said, "Yesterday I felt as if I had finally reached the top of a very high mountain. I was so happy about Judy's recovery. But I realize now we are only at a plateau. We've still got a long way to go before we ever reach the top."

Jim's words proved to be true.

15

February 10   When I came home for dinner Jerry was on the phone. He said, "A woman from the *New York Post* would like to talk to you about Judy."

The young woman asked many questions no one else had asked: Are you a religious family? Is Judy religious? Does she like poetry? What are her aims in life? I asked her to send me the story on Judy.

When Gene visited Judy in the morning she seemed very happy, but she was depressed and rather unkind to Jerry in the afternoon. He could not understand why she would treat him in such a manner. In the evening, when Jim and I were there, she was happy again. We thought Judy was taking her frustrations out on Jerry.

Later Judy read to us from *Diagnosis and Evaluation in Speech Pathology:* "An aphasic (brain-damaged) patient laughs or cries often, lapses into euphoria or withdraws into severe depression and despair. Behavior is largely a product of the drastically altered life experience, not merely the result of the damaged brain cells."

In the evening Carol Gevaart came to see Judy. It was the first time she had seen her. Judy did not seem to recognize her at first, but did after a short time. They had a short visit.

Carol offered to help us make arrangements to have Judy transferred to Madison. I knew we would not be able to see her as often in Madison, but Jim was eager to have her near him. I said to him, "Jim, will you be able to see Judy very often after we have her transferred to Madison? I know we won't be able to get there as often."

"I'll go see her every day," Jim assured me, "and if I can possibly go twice a day, I will."

"Thanks, Jim," I said. "It's good to know you'll be close to her and will be able to visit her often."

February 11   Many people were kind and generous to us during Judy's illness, and we were very grateful. Her fund had grown to $5400.

Our dentist, Dr. Duffy, came to the hospital to put a temporary cap on Judy's front tooth, with the promise of a permanent one to follow after her rehabilitation. He said, "I would like this to be my donation to Judy."

When the dentist told Judy to open her mouth wider, Gene said, "Now open your mouth nice and wide for the dentist, Judy." Since her accident Gene had been treating Judy like his "little girl" instead of a grown woman.

Judy became very indignant. She said, "Be quiet, Dad."

The dentist asked Judy if she would like to see how her tooth looked. He handed her a mirror. It was the first time she had looked at herself since her accident.

I was shocked when I looked in the mirror and saw a long thin face with big brown eyes surrounded by big bushy eye-

brows staring back at me. I couldn't believe it could be me.
I hardly looked at my tooth.

When Jerry went to see Judy in the afternoon she was very angry and wondered why no one had come up to visit yet that day, although Gene had been there most of the morning.

She was also upset because I'd taken the pictures down from her wall to show an interviewer. She said to Jerry, "Where are the pictures of the ones I love? I want them back on my wall where I can see them."

Jerry showed her a birthday card he had bought for her to give me the following day. She said, "I'd rather get one for her that I've picked out myself."

Jerry was hurt and upset. He didn't want to go back to the hospital in the evening. He felt that Judy was rejecting him.

When Jim and I went to see Judy I put the pictures back on the wall. She was happy to get them back and she seemed happy to see us. She told me about the nice card Jerry had brought her to give me. She said it had a nice verse.

Carol Gevaart made arrangements to have Judy transferred to the University of Wisconsin Neurological and Rehabilitation Center (N & R) in Madison. She came to the hospital in the evening and gave us the name of the neurologist to call. Then she said, "We have a large new motor home and my husband Jack said he would be glad to take off work and drive Judy to Madison so you won't have the expense of another ambulance trip. The motor home has a regular bed so Judy can lie down."

We appreciated Jack's generous offer. Carol gave me a picture of the motor home so I could show our doctor and get his approval to have Judy moved in it.

But we needed a paramedic to ride with Judy. Her skin

graft was still fragile and she had to be moved very carefully. That problem was solved when Jay Kurth, one of the paramedics who had volunteered his services on the ambulance run to O'Hare, heard what we were planning to do and volunteered to go with Judy to Madison.

February 12   The doctor approved Judy's mode of transportation. He called the neurologist in Madison and told him, "I'm sending you a healthy body."

The doctor told Gene he thought Judy would probably have an adult mind because he asked her the definition of "ecology" and "environment" and she told him right away.

Gene told Judy he loved her, as he had done all during her coma. Judy looked at him sadly and said, "I love you, and I tried so hard to tell you and Mom for such a long time."

With tears of happiness, Gene answered, "We knew you did, honey. We could tell how you felt without any words being spoken."

In the afternoon Judy was very tired so I stayed only a few minutes. As I was leaving I said, "Judy, I love you."

She looked at me and said, "I love you too, and you'll never know how much."

I answered, "I think I do know, honey. You've told us with your eyes."

That evening Lou Warner sent over our evening meal, a delicious ham dinner with a birthday cake. This was not the first time friends had prepared and delivered our dinner. It all started at a birthday dinner at the end of January when we gathered in the home of Mike and Donna Stluka with three other couples, Marcia and Arnie Lee, Joan and Deane Leavitt, and Dick and Dorothy Jentoft. Our group of five couples had celebrated our birthdays together for 15 years, and at the January gathering our friends decided to take turns bringing our evening meals so we could spend more time with Judy.

After our delicious birthday meal we went to the hospital so Judy could help celebrate my birthday. Judy was happy to see all of us. I opened my cards and gifts and everyone sang "Happy Birthday," including Judy. We were all off key, as usual, and Judy was the first to laugh about it. It all seemed so normal.

I thanked God for giving me the most precious birthday gift any mother could ever receive. My daughter came back to me a complete person. There were tears of joy in every eye. To this day I can't remember what any of my other gifts were.

We had just arrived home when the phone rang. It was Jim, calling to wish me a happy birthday and to say how bad he felt because he couldn't be with us. I told him how good Judy had been. As I was hanging up the receiver I thought I heard him say something, very softly.

February 13   Judy was in good spirits all day. The family decided to take me out for a birthday dinner in the evening, with Jim along as a guest. We didn't tell Judy about it because we didn't want her to feel left out.

Jim gave me a lovely birthday card with a warm message. He said, "I'm sorry I don't have any money to get you a present."

Jim was never one to show his feelings to anyone but Judy so the card meant a great deal to me—more than any present he could have given me. His card was signed, "A very special thank-you for everything. Jim."

When I asked him if he had said something to me before he hung up the night before, he said, "Yes. I said thanks."

"Thanks for what?" I asked.

"I wanted to thank you for never losing your faith or

giving up. I thought as long as you never gave up, how could I?"

He continued, "You know, I almost didn't go to that party where I met Judy. I feel I was supposed to go so I would be here to help Judy when she really needed me."

Jim and I met Gene, Janet, and Jerry at the Manor, a Beloit restaurant, to celebrate my birthday. For months our lives had revolved around Judy. We'd had so little time together. But we all knew there was still a long, long road ahead of us.

16

We received many cards and letters from friends and relatives and even from people we had never met telling us how thrilled they were that Judy had recovered. They wanted us to know they were rejoicing with us and praising God for the miracle.

One card I reread many times had a mustard seed in plastic casing. It said: "If you have faith no bigger even than a mustard seed . . . nothing will prove impossible for you."

A young woman from Denver sent this letter of encouragement:

Today I read in the *Denver Post* of your recovery from the coma. I share in your parents' joy, as I personally know what you've all been through. My two brothers and I were overcome by carbon monoxide in Mexico in 1971. I was in a coma only two days, but my brothers were comatose for seven and forty-two days respectively. They both were on the brink of death at times, but they recovered fully. One is now finishing his M.A. degree and the other is pursuing a B.A.

Judith Ann had remained in Spain to teach. Her parents kept her informed of Judy's condition, and when she heard about Judy's recovery, she wrote:

> Hello to you, honey! You can't imagine how I felt when I heard the news of your miraculous recovery! Thanks be to God, who listened and heard our prayers! I've never prayed so hard in my life.
>
> I am ecstatic and overjoyed that you have recovered. I screamed when I learned that you were talking. I told many of my students about this miracle. It's upsetting to me that I can't be there. I wanted to catch the first plane out of here to come and see you and talk to you and hug you! I dreamed just about every night that you'd come out of that sleep. I wanted it and prayed for it *so* hard. I just cry for joy when I think about the powers of God, and he used them to heal you!!
>
> For a while here in Madrid, life was really a drag for me. I didn't have my heart in my job, I had no ambition, and I wanted to go home. But things are much better now, and especially since you have recovered! I like my classes and my students are fun and neat. Briam Institute is a nice place to work.
>
> I won't be coming home until the beginning of June so I won't be able to see you for another three and a half months. That's a long time, isn't it? I wish I could talk to you! You are a real celebrity around the Beloit area I hear: radio, television, and newspapers!!
>
> You have been and still are in my thoughts and prayers. God bless you and your family.

So many people took the time to tell us they cared. We have found the world to be full of wonderful, caring people. Our whole family felt truly blessed.

17

February 14   From her students Judy received valentines and a letter:

Dear Judy Steuck, How have you been lately. I hope you are fine because I am fine myself. Judy I am glad you are feeling better. I heard that you like my get well card. Judy I hope you get much better because I miss you very much and I will think of you all the time when I am sleeping and I will think of you every day and night. I will have a lot of dreams of you and you have a lot of dreams of me Lorraine. Judy I wish you a very Happy Valentine and I will sent you a valentine card but I will be late. Judy I will tell you my birthday and my birthday is February 19th and that's Thursday next week. Judy, Tammy misses you very much and she wishes you a Happy Valentine too. Judy, Tammy will write to you when she has time. Judy I heard that Mrs. Gevaart goes up and sees you every night and Mrs. Gevaart says hi from all of us in class. Judy do you miss me while you are in the hospital? Judy I know you can't write to me yet but when you feel OK to write me you can. Judy I saw your picter in the newspaper when you woke up. Judy your hair got longer since you seen me, Kevin and Tammy. Judy I wish you were our Speech teacher again because we miss you alot. Judy please write to me and send me a picter of you and I will send you a picter of me and Tammy. Judy that's all I have to say to you. Your friend always, Lorraine. P.S. Judy I hope you feel much better.

Jim came to see Judy in the afternoon and brought her a valentine and a dozen pink rose buds. She seemed pleased with the roses, but she said, "I thought you would bring me something in a little square box" (an engagement ring).

Judy was becoming obsessed with wanting to get married as soon as possible. It was one of the things she thought of all during her coma, which gave her the tremendous will to live. Getting married now was all she talked about to the nurses.

Judy started to have hallucinations. She said some "flakey" things. Whenever Jim heard her speak this way he would say, "O.K., Judy, stop talking goofy and be serious with me," and she would be more rational.

We now had permission to try to give her a few solid foods. When I started to put a spoonful of pudding to her mouth she said, "Mother, why is that dog sitting on your hand?"

I tried not to act surprised at her question, so I said, "Judy, there can't be a dog on my hand. Animals aren't allowed in the hospital."

She said, "Oh," and seemed satisfied with my answer.

When Gene was with her she said, "Dad, why have you got one black hand and one white hand?" Gene tried to pass it off, but it really bothered him when she said such strange things.

Judy started talking about our neighbor's youngest daughter who had recently married. She laughed and said, "I can't imagine Katy getting married when she is only in grade school." We thought Judy was really mixed up and we couldn't understand why she kept talking about our neighbors.

Judy had a nurse she called Robert, although that wasn't his name. Then one day Janet said, "Did you happen to notice how much the male nurse, Minor, looks exactly like our neighbor's son, Robert?" We suddenly realized Judy had thought she was talking to Robert. She really had not been hallucinating at all.

I was sure Minor was Robert. He talked about going to see his aunt and he could never understand how I knew she lived in Chicago. He also said his younger sister had gotten married. I thought it must be Katy who had gotten married and I thought it was so funny because she was only in grade school. I had to tell my family the news.

Judy also had a nurse named Rosemary. Judy had called her Thyme since she had been introduced to her. Rosemary could never understand why.

When I was introduced to Rosemary, her name reminded me of the song that says, "Are you going to Scarborough Fair, parsley, sage, rosemary, and thyme." So I just called her Thyme whenever I saw her.

Pastor Nycklemoe came to the hospital to give Judy Communion. I said to Judy, "Pastor Nycklemoe is here and wonders if you would like to take Communion. Would you like to take it now?"

Judy surprised us both with her answer. She said, "No, I don't think I'd better take it right now because I just had some grapefruit and I don't think the wine and grapefruit juice would mix very well."

I was embarrassed, but Pastor Nycklemoe said, "I'll come back another time, Judy."

Gene and I decided to sponsor a radio broadcast of our church service to show our gratitude for the congregation's many prayers and donations. I decided to sign up for February 8, thinking Judy might be a little better by that time. By a very strange coincidence it was the day *after* Judy had come out of her coma, a very appropriate time to sponsor the broadcast and thank all the people in the area.

Broadcast sponsors are listed in the church paper. I showed Judy the announcement in the church bulletin saying, "The Voice of Our Savior's: Mr. & Mrs. Eugene Steuck in gratitude to God and thankfulness to the congregation and community for their prayers for Judy."

With a strange look in her eyes, Judy said loudly, "Prayers for Judy? What is this?"

I realized that for some reason she could not comprehend. Nothing more was mentioned.

I was so confused because of this terrible thing that had happened to me. I didn't know what I could have done in Europe that could have been so bad that God would want to punish me this way. Did I do something that would make Jim ashamed of me? Why did Pastor Nycklemoe want me to take Communion? Did he want me to repent because I had been such a sinner? Why was my name in the church bulletin? Did they all know something about me I didn't know?

Although I have always had strong faith in God, I had no religious thoughts during my coma. I was too confused about what had happened to me and why I could not speak or move. I could not understand why Jim was being so kind to me. I thought I must have done something very bad.

It never got through to me when I was in my coma what had happened to me. I did not understand the gravity of the situation.

Jerry told Judy that Gene had given a radio interview

regarding her recovery. Again a very strange look came into her eyes as she asked, "What?"

Having read Judy's eyes for so long, we realized she should not be told anything about the widespread news reports about her recovery. We tried to make light of Jerry's comment and never allowed anyone to tell her anything about the news media.

We thought Judy might have to undergo psychiatric treatment when she got to Madison. She'd had a horrifying experience, and we decided it would be best to let a professional handle her problems.

When Jerry mentioned WBBM it just didn't make any sense to me that they would be interested in my illness. I didn't know why there should be so much excitement about me. It wasn't until I got to Madison and had undergone diagnostic testing that I realized I'd had carbon monoxide poisoning from the space heater in Spain.

February 15   Judy remained in ICU but a few friends were allowed to see her. She became quite tired from the extra company and wanted to rest. Jim said, "I'll stay here with Judy and you can get something to eat if you like. I'm not very hungry."

I went home for a short time, then returned at dusk with Jerry. All was dark in Judy's room. Jim was sitting next to her in a chair, holding her hand. They had fallen asleep and they looked so content I didn't want to disturb them, but Jerry was eager to talk to Judy again so we stayed in the room and they woke up.

We all stayed until 9:00. We were looking ahead to tomorrow, the big day Judy would be transferred to Madison.

18

February 16   As we prepared to leave, Doris, one of Judy's nurses, came rushing up to Judy and said she wanted to kiss her goodbye. Then she asked, "Is Judy really going to get married February 27th?"

"No," I answered, smiling. "She just has her time mixed up. She won't be able to get married for some time yet."

"I didn't see how she possibly could," Doris said in a very relieved voice, "but she did say she was going to get married then."

Just as we were leaving the hospital, a young woman came over and said, "I've just got to see her. I can't believe this. I'm the one who gave Judy her EEG the end of January."

She looked on in amazement as Judy was put into the motor home.

When Judy arrived at Madison she was down from her

usual weight of 120 to 80 pounds. She could barely lift her head off the pillow and was unable to even hold a pencil in her hand.

After she was admitted, we had a conference with the neurologist. He said, "I suppose you are aware that in most cases of carbon monoxide poisoning the victims are dead in two weeks from cerebral edema. But as long as Judy has progressed this far, we will be testing her thoroughly all week to determine what we can expect of her."

After Judy was settled in her room, I called Jim and held the phone so Judy could talk to him. It was the first time she had talked to anyone on the phone since her accident. Jim said later she had repeated everything he said, trying to comprehend. He told her he would come to see her shortly.

The physical therapist and occupational therapist came in to meet Judy and check out her condition.

Sharon, the PT, came into my room with Karen, the OT, and asked me to roll over onto my right side. I said, "I can't turn over. It hurts too much." Something really hurt in my left leg. I didn't know what it was, but it felt like a rock in the way.

Dad said, "Come on, Judy, you can do it. It might hurt, but remember what we told you. To walk again you're probably going to have to do some things that will hurt."

Judy was unable to ring for the nurses, so they connected a little pillow to the light. All she had to do was lay her arm on the pillow and the light would go on so the nurses would know she needed them.

When we left in the evening it was such a comfort to see Jim sitting by Judy's bed and to know he would be with her when we were unable to be there.

That night Judy cried most of the night. Jackie, a young nurse who was Judy's age, heard her crying and couldn't figure out what might be wrong with her. She checked her chart. After reading page after page of what had happened to Judy, she felt deeply sorry for her. She told me later she was tempted to do anything she could to help Judy, but she had been trained not to help the patient any more than necessary. She said she had to watch herself so she would not wait on Judy too much.

I could see the capitol from my window. I knew I was back in Madison, where I'd had so much fun. I also knew it was going to take a long time for me to get better. I didn't see how I would be able to get married by the end of February, but I was really going to try to get better so I could.

Judy was very disoriented and kept asking Gene if he had done his Christmas shopping for me. She couldn't believe she'd missed Christmas, because she had eagerly looked forward to it when she was in Europe.

Judy's friend Jolene brought her a calendar and we showed her what day it was and marked off each day as it passed.

February 17   Since Judy had become upset when the news media was mentioned, we were concerned that no one come to ask her questions, which might confuse her and hamper her progress. We were aware of all the press calls to the hospital, reporters wanting to "talk to the coma patient."

I asked Jim if he would make a list of people who would be allowed to see her. He also cautioned all his friends about what not to say in her presence. He hurriedly made out a list and gave it to the nurse, but in his haste he inadvertently left out some familiar names.

My brother Harold came up to see Judy a short time later and was not allowed into her room because his name was not on the list. Although he told the doctors and nurses he had helped bring Judy home from Spain, they wouldn't allow him to see her because they were convinced he was from the press. Harold finally called Gene to request his permission, and then his name was added to the list.

The same thing happened to our youth pastor, Paul Walker. He also had to call Gene for permission to see Judy.

I felt secure in the knowledge that Judy was being closely watched. Her room was next to the nurses' station.

February 18   I felt lost because I was unable to see Judy every day. Going to the hospital had become a daily routine.

I came home at noon. As I was eating my lunch, the telephone rang. I was surprised when the operator asked if I would accept a collect call from Judy Steuck. A weak little voice said, "Hi, Mom. When are you going to come up and see me again? I miss you and Dad."

"I'll be up tonight after work," I assured her, "but Dad can't come until Sunday." She seemed satisfied.

I knew Judy was unable to hold the phone in her hand or dial the number. When I saw her I asked how she'd made the call. Although she had not called our number since Thanksgiving Day, collect from Denmark, she remembered it and told the nurse, who dialed for her and held the phone so she could talk.

I received a collect call from Judy every day for the next three weeks during my lunch break at home. It was good to hear her voice and to know she could finally talk any time she wished.

February 19   I received a letter from the *New York Post*. Eagerly, I tore it open, anticipating a nice story about Judy. The article was dated February 11, four days after she had come out of her coma. I was shocked at the headline: "MD Doubts Coma Miracle."

The reporter had asked me for one doctor's name. She quoted the doctor's words in a telephone conversation:

> "There is no miracle. The girl is still in a coma. She started making gurgling noises about four or five days ago. It could be they thought she was talking, but from a medical point of view. . . ."
>
> Like Karen Quinlan, the headline-making New Jersey coma victim, she suffers from "decerebrate rigidity." She does not lie in a Sleeping Beauty repose, but has frequent spasms described as "typical of someone who does not have high neurological functioning."
>
> "She is feeble, all crippled," said the doctor, "making unintelligible sounds."
>
> Miss Steuck, her physician said, cannot be described as being in a "persistent vegetative state" because "there has been a slight improvement" in her brain waves.

The article concluded, "He summed up the case starkly as that of a girl in a coma induced by very severe brain damage."

I was shocked. I could not understand why our doctor had still thought Judy was in a coma. He had seen her when everyone else was aware she was out of it.

When the doctor came into my room (Monday, February 9) and started talking loudly, as if I could not hear, and treated me as if I were a baby, I knew he did not believe I was normal. I knew he was testing me. It made me angry and I wouldn't do anything he asked me to do, not even drink out of a straw, although I had drunk from one the night before, when Jim and my mom were there. When the doctor came back another day and asked me to give him the definition of "ecology" and "environment" I told him, because he treated me as an adult.

The medical student who had interpreted to the Spanish

doctor was with our doctor. He asked me some questions in
Spanish. He seemed surprised when I answered him in
Spanish because I told him I had not taken any Spanish
classes since fifth and sixth grade.

February 20   The diagnostic testing was completed.
Judy was diagnosed as a "spastic quadriplegic" (no control of upper and lower extremities) with an I.Q. of 85,
somewhat above borderline retardation. (A few weeks
later it was 110.) But Judy retained a fantastic memory.

X rays revealed a calcium deposit in her left hip which
kept it from rotating. This was from inactivity while she
was in her coma. Because of burns on her right side, she
had almost always lain on her left side facing the window,
except when she was flipped on her stomach in the morning and afternoon. Her right side had the badly burned
area and it was uncomfortable for her to lie on it.

The calcium was first believed to be the result of a bruise
Judy had received while skiing in Europe. But the bruise
on her leg was not on the muscle on her left hip where
the calcium deposit had formed.

The neurologists had a conference with Judy and told her
exactly what her condition was and what they planned to
do for her rehabilitation. They knew she had graduated
from the university as a speech pathologist and how intel-
ligent she was before her accident. They hoped, with the
right kind of stimuli and rehabilitation, she would progress
to her former I.Q.

They planned to get her into a wheelchair. Then they
were going to teach her how to use her hands so she could
feed herself and write. The third step would be to teach
her to take care of her personal bodily needs. Then they
were going to teach her how to walk again.

The neurologist asked Judy, "How would you like to leave this place—in a wheelchair or walking?"

Judy answered, determinedly, "I'm going to *walk* out of this place."

The neurologist answered, "Then you have a lot of work ahead of you."

19

When we arrived in Judy's room she exclaimed, "You know what? I was overcome by carbon monoxide poisoning from the heater in my room in Spain."

I answered, "Yes, Judy, I know." How well we all knew!

I had heard people talking about carbon monoxide poisoning, but I didn't understand how I could have been overcome by it. I didn't think about the space heater in my room in Spain.

When I saw Jim I said, "I was poisoned by carbon monoxide fumes from the space heater in my room in Spain." I wondered why he didn't act surprised. I thought I was telling him something he didn't know.

Testing showed Judy's calculations to be off. Jim had realized Judy couldn't add numbers. He decided to bring his calculator to her room. He told her he wanted her help for papers he had to get done for his class. He asked her to do some easy calculations to see if it would help her get numbers correct in her mind again. She was happy for the chance to help him. Soon she could add simple figures, although not quickly.

When she opened her cards, she never seemed very excited when a 10- or 20-dollar bill was enclosed. Finally I realized she didn't know the amount of the bill. But soon she relearned the value of 10- and 20-dollar bills, and she still prefers them to ones and fives.

Testing also revealed Judy had received in her left ear a mild hearing loss in the high frequency range which she will always have. The doctors said it is not uncommon for CO victims to completely lose their hearing.

Judy became very depressed while going through all the diagnostic testing. She would say, "I am so tired of everyone asking me, 'Do you remember this?' or 'Do you remember that?'" When Jim came to see her, it sometimes took him an hour or two to get her to perk up.

I had become aware of how Judy felt about being asked that question almost from the beginning and never asked her if she remembered something. If she wanted to tell me about an incident she remembered, she would just casually mention it.

I had always thought Judy might have to undergo psychiatric treatment when she became aware of what had happened to her, but now it didn't seem necessary. With Jim faithfully helping her out of her depressed moods, she didn't go into a deeper depression.

Judy did have a complete mental block, however, about what had happened to her during her coma. It was not until June 22, 1976, when the humming of the fan in her window reminded her of the heater in her room in Spain, that she started to remember. Then she continued to regain her memory very slowly.

February 22   They put Judy in a wheelchair for the first time. It was good to see her sitting up, although her body looked like a wet dishcloth.

I was transferred from my bed to the wheelchair with the assistance of three nurses. I couldn't do anything to help. When I was in the chair, I just dangled. I couldn't support myself with my legs at all. My arms had no strength and I couldn't give myself a boost up in the chair. I could hardly believe I was so weak. I felt like a Raggedy Ann doll. I just hung, with no control over any of my muscles.

Jim was more successful than anyone else in getting Judy to try to do things. He said, "I never want to hear you say 'I can't do it,' because I know you can." Judy believed him and she always tried.

Jim's parents and his sisters came to see Judy quite often. They all treated Judy as if she were already a member of the family.

Gene worked until 7:00 P.M. every day except Wednesday and Sunday. Since it was 60 miles to Madison and visiting hours were only until 8:00 P.M., he could see Judy only twice a week, but I usually went to see her every day except Tuesday and Thursday. She asked me to be there by 5:00 so I could feed her because she didn't like to make the nurses do it. She knew they were very busy.

Judy's friend Joy Bjorklund lives in Madison. She usually tried to help feed Judy when I couldn't be there. She also did many other little things Judy needed done.

February 23   Judy started her occupational and physical therapy but needed no speech therapy.

When Judy first tried to feed herself she couldn't direct the spoon to her mouth. When she did get it to her mouth, she usually spilled most of her food first. But she just laughed. In a short time she was feeding herself very well.

Gradually she learned to grasp a pencil. Then she started learning to write all over again. Her writing was very bad at first, but by the first week in March we could read her writing quite easily.

I had been told that when coma patients learn to write again their penmanship is usually entirely different from what it had been, but Judy's is almost like it was.

The catheter was removed. Everyone was pleased with Judy's rapid recovery. She was determined to get well again —so she could get married.

February 25   Judy kept her usual sense of humor. When Jim asked Gene and me to dinner at his apartment, Judy asked, "Will you take something to Jim from me?"

I said, "Sure, what do you want me to take?"

With a glint in her eyes, she said, "A can of Comet. If Jim's apartment looks anything like it used to, he'll need it."

Judy warned everyone not to come to see her after she got out of physical therapy because she was always so worn out.

My PT's name was Sharon. She was a very good physical therapist, but she was only human and didn't enjoy seeing me in real pain and sometimes in tears.

One day the doctor came to watch my therapy, to see if I was stretching the muscle in my left hip sufficiently. He thought the calcium deposit should be worn down further by now, which would enable me to walk.

I was afraid of him and wanted to think of some excuse to get out of there. Sharon told me not to be afraid because he would probably just watch.

He did watch for a while, but he decided I needed someone to make it harder for me, so he proceeded to work with me.

I was in pain and sobbing. I didn't believe any human could be so cruel. It seemed to me he was working me harder than a slave driver would.

He made me get down on my hands and knees and crawl to the edge of the mat. I could hardly support my weight with my arms. My legs seemed to be stuck and I was afraid of falling off the edge. I sobbed and told him I was in pain but he seemed to have no mercy.

He continued in his loud voice, "Come on, you can do it. You have to if you ever expect to chalk down that bone."

I was bitter toward him and I told everyone how terrible he was to me. I even told Pastor Nycklemoe. I thought that doctor was the meanest man in the world when he came to see me right after I came from PT that day.

Later I realized the doctor was trying to get me so mad that I would want to work even harder so I could get out of there.

When I was in Madison visiting Judy, a man called our house and talked to Gene. He told him his 13-year-old son had been in a coma since Christmas and the doctors gave no hope for his recovery. He had heard about Judy through a mutual friend and said he had to talk to someone who would understand what he was going through.

Gene talked to him about half an hour and encouraged him and told him to keep praying and never lose his faith.

Gene told me about his conversation. "We must be sure to pray for the family tonight." And we continued to remember the family in our prayers.

About a month later, the man called back. He said, "I just had to call and thank your husband for talking with me when my son was in a coma. He gave me much hope and renewed my faith. Our prayers were answered—my son came out of his coma about two weeks ago. He's doing well, although he'll need a lot of therapy. He doesn't have his complete memory, but we're thankful God gave him back to us."

I called Gene at work to tell him the wonderful news. We both rejoiced because another family had felt God's healing hand.

20

March 7   Our family and Jim were talking with Judy when she casually said, "Did my grades come from my semester in Europe? I know I must have gotten 4.0 because I studied so hard."

I saw the look of surprise on the faces of Jerry, Jim, and Gene. I didn't want to upset her, so I said, "I can't remember. I think they did but I'll have to check."

When Judy was in her coma we had received her grades —two As and two Fs. I was so disappointed. I thought, "Judy is in a coma after going to Europe to study and she ends up with two Fs." The whole thing seemed like such a waste.

I knew Gene would be bothered terribly if he knew so I hid the grade report, thinking I would be the only one who would have to know what her grades were.

After we left Judy's room, Jerry said excitedly, "I didn't

think Judy could have gotten two Fs. She's not that kind of student. I told Jim and he didn't think so either."

"How did you find out about her grades?" I asked, surprised.

He said, "I saw where you put them and looked to see what they were. I couldn't believe it. Then I told Jim about them, but I didn't want to say anything to you or Dad." Gene had also seen them but had said nothing, each of us trying to spare the others.

I immediately wrote a letter to the professor who had been in charge of their semester in Europe and asked him if he would check into it. He called me right after receiving my letter and said he knew Judy had done well in her studies. He thought there must have been an error, and he assured me he would look into it.

A short time later he called me again and said Judy had indeed received 4.0, and her records would be corrected.

We were all happy to hear this and I told Judy, knowing it would not disturb her as long as everything had been corrected.

Judy was progressing very well with her therapy, although it was painful for her. The neurologists were happy with her progress. One of them came into her room and said, "Judy, can you read this card to me? You don't have to read anything that might be personal."

Judy took the card and read it to him. She also read the entire personal message.

He looked surprised and asked, "Would you like to read a novel?"

March 10   When Gene and I arrived, Judy, dressed in new clothes, was sitting in the hall in a chair eating her evening meal. She was by the elevator waiting for us to come so she could tell us her good news immediately.

With a big smile, she said, "My PT had me walk by the parallel bars today. She told me to try to take two or three steps, but I walked the whole length of the bars."

Gene and I beamed with happiness. I said, "Judy, that's just great. It won't be long now and you'll be using a walker."

The doctors and nurses were also all smiles.

After Judy went to her room, I said to the neurologist, "When Judy was in her coma Gene promised her a trip to Florida when she was better. Do you think it would be all right for me to make reservations for the end of May?"

He answered, "That would really do her a lot of good. She should be walking quite well by then. But if she isn't you could always rent a wheelchair there and take her around."

Then the neurologist tried to explain in everyday language what had happened to Judy's brain. He said, "She was very lucky they gave her such good treatment when she arrived at the hospital in Spain. That prevented her from getting cerebral edema, which kills most CO patients."

He drew me a simple picture and explained, "We don't understand how it happened, but no cells were damaged in the gray matter in Judy's brain. That would have caused permanent brain damage. Damage was done just to the brain cells in the white matter. These cells undergo a type of regeneration and will keep regenerating until Judy is completely normal, which could take a couple of years."

When I told him our Beloit doctor had said we should build a room on our house, he said, "Well, I'm going to tell you to plan on making a wedding dress and being a grandmother."

When they heard that encouraging statement, Judy and Jim made plans to be married October 29, thinking this would give Judy enough time to get her strength back and for her to be able to walk well again.

Many people offered to take me to Madison, knowing how tired I must be from driving back and forth all the time. All who had prayed for Judy were eager to see her and tell her how happy they were with her progress. Usually on Friday, when I was the most exhausted, I would accept an offer and get a ride to Madison.

One Monday I received a copy of a letter LaVaughn Kunny, the woman who had driven me to Madison on Friday, had written to her family:

A week ago Friday I picked up Jeanine Steuck, the mother of Clare's close high school friend Judy, who had been overcome with CO in a rooming house in Madrid. Judy was in a coma, in a fetal position, for 52 days. I was so happy to see her and observe her response to all of us. This seems to be a miracle, as the family was offered no encouragement. They didn't miss seeing her one day, and her boyfriend came from Madison and was with her much of the time. They talked to her, read to her, held her hand . . . and would not let her leave them. It seemed to me they loved her back to life! What a wonderful thing for all of us who know her to witness. It must be a modern day miracle.

21

March 14   The people from our congregation wanted to do something more to help with Judy's mounting expenses, so they came up with the idea of a benefit.

Our church sponsored a benefit concert by the Rockford Kantorei, a well-known boys' choir. The church was filled with many people I had never seen before. The freewill offering of $1000 was matched by a grant from the Aid Association for Lutherans of Appleton, Wisconsin. Combined with previous donations, Judy's fund reached a total of $7400, a tremendous help towards the expenses of $14,000. We have always been very grateful to everyone who helped. I felt it was God's way of helping us with expenses.

The next day Judy asked, "Did anyone come to the concert yesterday?"

"They sure did," I answered. "The church was full."

"It was—really?" Judy couldn't believe so many people

would come to a benefit for her. She didn't realize how many people had shown us how much they cared since her accident.

With Judy's continued improvement, visiting restrictions were lifted. All her friends from Madison, Beloit, and elsewhere came to see her. The nurses said they had never seen anyone have so many visitors.

Judy now became self-conscious about her appearance. She hid her "greasy" hair under a scarf and complained about her "zits." It was apparent that she was getting better.

March 20   Judy started to use the walker. With Jim's help and encouragement, she learned quickly.

Yet she had some depressing days, especially when Jim took her outside to use her walker.

When I saw all those young girls go by, carefree and walking normally, I would look at Jim and think, "I know he must feel very embarrassed walking with me in my condition when all the other girls are walking normally and look so young and healthy."

I thought he was looking at them with lust in his eyes. I became very jealous whenever I thought he was looking at any pretty girls.

When I was visiting Judy she started to cry. "What if they can't get this calcium worn down so I can walk again? Won't they ever let me out of here?"

"Of course they will," I assured her, "but you have to work very hard in therapy no matter how much it hurts you so you can get that calcium chalked down."

Judy told her PT what I had said the day before. Sharon said, "Your mother is right, Judy. You have to work very hard, even though it may hurt a lot. Is your mother a teacher?"

"No, she's just a mother."

Being "just a mother," it is easy to love too much and become too protective and possessive. When Judy was in her coma I had become accustomed to doing everything for her. I had opened all her mail and decided what should be read to her and what could be said in her presence. I made all her decisions. I continued to do many things she was now capable of doing herself.

A letter came from an honor sorority asking Judy if she would like to become a member. They wanted an immediate reply, so I took it to the hospital and asked Jim what he thought. He surprised me by saying, "Why don't you ask Judy?"

I had been so accustomed to making Judy's decisions for her that I didn't even think of asking her. Jim made me see the light. From then on I tried to treat her like a responsible adult.

March 29   Judy had always enjoyed hearing Janet's church choir, the Disciples of Distinction. Since she couldn't go to church to hear them sing, Pastor Walker, the director, decided to bring the choir to her.

Permission was granted for the choir to sing in the occupational therapy room. They sang all the lively and spirited songs Janet knew Judy would love to hear. Judy was all smiles. It was difficult to tell who enjoyed the singing more —the patients or the staff.

The next day Judy slowly typed a thank-you letter to Pastor Walker.

April 2   Judy started to walk with a cane. She still had a bad limp, but we hoped that would go away after the calcium wore down.

Judy called and said, "Do you know what I did? I walked all the way down the long hall to the waiting room and back. Jim was beside me urging me on."

"That's great, Judy," I said. "That was really a long way to go. Plan on walking for us when we come to see you tomorrow."

But when we went to see her the next day, Jim was not there. He seemed to be the only one who could get Judy walking. She walked just a short distance for us.

April 4   Judy was allowed to leave the hospital for short intervals. We would take her for rides around Madison and stop at a park. Jim would take her out at night to visit friends. It was good therapy for her, but she was becoming eager to go home.

Finally the doctor gave us permission to take Judy home for Easter weekend. In her excitement Judy said, "Will you do something for me as soon as you get home? Will you make an appointment at the beauty parlor for me to get my hair cut? I just can't stand this straggly hair."

"Sure," I answered. "Is there anything else?"

"Do you think I could stop on the way home and get a new Easter dress?"

"Of course," I answered thankfully, because the conversation seemed so very normal. Judy was acting like her old self again.

April 17   Janet and I arrived at 10:00 A.M. Judy was impatiently awaiting our arrival. She could hardly wait to get into the car and head for home.

But the shopping center was the first stop. We were there only for a short time and Judy became very tired. She settled for a pair of slacks and a top.

At home Gene was anxiously awaiting Judy's return. It was a wonderful sight when she walked through our doorway again after an absence of seven and a half months.

She had a happy reunion with Pepper, our dog of 13

years. Then she went upstairs to her room. Before leaving for Europe she had told Janet, "Be sure to water all my plants and take very good care of them so none of them die before I get home." Janet didn't have a green thumb, but she watered those plants faithfully. Judy was delighted.

Happily, she opened all the packages she had sent home from Europe. Each one seemed to bring back some happy memory.

Judy's haircut did wonders for her morale and she looked very pretty. She came back home to go through her luggage and found all the Christmas gifts she had bought in Europe. She asked me for some brown paper bags. She put her gifts in them and wrote the recipient's name on each bag. We put them by the artificial Christmas tree I had set up on the porch.

Jim came down to help us celebrate Judy's special Christmas. He was pleased with her new haircut and outfit and told her how lovely she looked.

My parents had returned from Florida and were invited to our celebration.

We gave Judy all the packages we had wrapped and kept for her almost four months. She received a book from Jerry titled *Thoreau Revisited: Diary of a Country Year*. The inscription said: "Judy—Merry Christmas 1975. The ultimate aim of human life (which is also a definition of sainthood) is to live every day in the physical universe as though it were our first day, and to live every day in the moral universe as though it were our last day. I know of no better example of this than you. Love, Jerry."

With thankful hearts we opened our gifts from Judy. For months we hadn't known if she would ever be able to tell us for whom she intended her gifts. As I looked around the circle of radiant faces, I thought, "Praise God from whom all blessings flow."

We had our special Christmas dinner and offered a prayer of thanksgiving for Judy's presence with us again.

April 18   Judy hadn't outgrown childhood traditions. She and Janet searched the house for the Easter baskets the Easter bunny had left during the night. But Pepper had gotten there first. He sampled a big chocolate egg from each one.

We went to church for Easter breakfast prepared by the High League. Judy was greeted by many people who hugged her as tears rolled down their cheeks. They called her a living miracle.

As we sat in church, with Judy between us, I was overwhelmed by God's great love. Yesterday we had celebrated the birthday of Christ. Today we were celebrating the day he arose from the dead. "For God so loved the world that he gave his only begotten Son." He showed his love by giving Jesus to live and die for us. His love lives on in the hearts of people who know him and believe in him. Judy and our whole family had felt the love of God through the love of people all over the country who prayed for Judy and helped us in many ways.

That evening we took Judy back. It was hard to leave her in Madison, but the neurologists wanted to do a week of final testing.

April 21   When I answered the phone, Judy was crying as she told me her news, both good and bad. The good news was that she would be discharged on Saturday. The bad news was that she would have to undergo surgery on her hip. There was no other way to remove the calcium. Surgery could not be done before six months because they wanted to make sure the calcium was localized. Judy and Jim would have to postpone their wedding plans.

22

Jim was disappointed when Judy told him she would have to undergo an operation on her hip. "All that hard work for nothing," he groaned, "and you'll have such an awful scar."

Jim had received several job offers. He and Judy decided on a company which had plants located around the United States and in Caracas, Venezuela. He accepted a position but would not know until June where he would be sent.

They set their wedding for December 27. Judy had made arrangements to have her October wedding reception at the country club and checked to see if she could get it postponed to December 27, but all dates were booked up for December. The club always closed from January till March, but because Judy had worked there as a waitress the previous summer the manager offered to stay open so she could have her reception there January 8.

Judy decided to go ahead and make the reservations without consulting Jim. He was upset when he found they could not be married in December. A wedding two days after Christmas during his vacation visit home would have been more convenient. The changed date seemed to be one more obstacle in the way of their marriage, but Jim went along with the new plans.

Judy had to go to the Beloit hospital three times a week for physical therapy to build up her unused muscles. I helped her for more than an hour each evening with the therapy exercises because she was unable to do them by herself.

She walked without a cane, but with a very slow shuffle. Her left hip couldn't rotate because of the calcium. She was still quite spastic and without help was unable to walk off curbs or up and down stairways that had no railings. When she would lie on the lounge in the backyard, she would be unable to get up without assistance. Many times I looked out the window to see her trying in vain to get off the lounge and yelling, "Mom, come and help me!"

It was like having a little child again. She needed help in the shower and in the bathroom and with many other things. She would laugh uncontrollably for no reason, usually when she was in church or at other serious times. When I mentioned this to Jim he said, "Judy has always been a very happy person." He seemed reluctant to acknowledge that Judy had a long way to go before she would be herself again.

My mother always came to stay with Judy in the morning until I came home from work. She also helped Judy with her therapy exercises.

Judy enjoyed Jim's parents and always looked forward to her frequent weekend visits at their home. The trips to Madison broke up the monotony of her daily life.

Judy taught a fourth-grade vacation Bible school class at our church every evening for one week. She enjoyed it and did quite well. She usually got a ride home with a friend, but one evening she didn't come home at her usual time. I started to worry, but I kept telling myself I was being too protective. She finally arrived home about an hour later.

She had asked a little boy to take her hand and help her down the church steps because there were no railings. He put his hand out to her, but he offered little support. They both fell down the steps and Judy hit her head on the door. Luckily no one was hurt. Judy tried to be more cautious from then on—and I became even more protective.

May 21   Judy was interviewed by a local radio station. She was unable to say much about her coma because she still had a mental block and remembered nothing. But she thanked all the people in the area for all they had done for us.

The interview was to be aired May 24, because Judy wanted first to thank her congregation family on May 23.

May 23   Judy spoke at all three of our morning church services. I took her to each one but sat in the chapel as she gave her little thank-you speech. I saw tears in the eyes of many as she gave this message:

> This is my opportunity to thank all of you for the help and gifts you gave to my parents and to me, and for all the prayers for my recovery during my recent coma and weeks of convalescing.
>
> It is not possible for me to name all who have done so much, but I feel I must mention Pastor Glenn Nycklemoe for all his help and prayers, Pastor Walker and the Disciples of Distinction for coming to the Madison hospital and singing to us, and my uncle Harold Cole for accompanying my mother to Spain and helping with all the arrangements to return me

to Beloit. A special thank-you to my mother and father, Mr. and Mrs. Eugene Steuck, my sister Janet and brother Jerry, and my finance Jim. God bless you all.

A theological student who heard Judy's remarks commented, "We didn't need any other message for the day. Judy was the only message necessary."

Judy wrote the guest editorial for the Monday, May 24 edition of our Beloit newspaper and gave essentially the same message.

May 24   The time finally arrived for our trip to Florida. It was Janet's first plane flight and she enjoyed it immensely. Judy, Janet, and I landed in Orlando and took a bus to our hotel. We had a lovely room overlooking the pool. It was very restful, just right for Judy's rehabilitation. If we wanted a little noise and excitement, we just had to hop on the Monorail and go to Disney World.

Since Judy walked so slowly, we rented a wheelchair and went out to have breakfast. While we were inside it started to rain and the wheelchair was drenched. We decided to go back to the hotel.

The next day the weather was good, so we headed for Disney World. Judy found out what it's like to be in a wheelchair in a big crowd and to have people stare at her, but it didn't bother her too much. Janet had a different reaction. Whenever she saw someone stare at Judy she would say, "People sure like to stare, don't they?" After hearing Janet's comment the people would quickly glance away.

We went back the next day. As we got on the train that circles around Disney World, it started to sprinkle. After we left the station it started to pour and we got soaked. There was nothing we could do about it so we just laughed and had a good time.

Janet gave Judy a wild ride back to the Monorail. People jumped out of her way as she ran with the wheelchair. We were anxious to get back to our hotel so Judy could take a hot bath and not get chilled. We were concerned about her getting pneumonia again.

Judy and Janet still talk about their many happy memories of the trip. It helped boost Judy's morale.

Judith Ann finally returned to Beloit and came over to see Judy. As I met her at the door, she said, "I've dreamed about this meeting with Judy many times. I'm eager to see her."

Judy and Judith Ann had a happy reunion and went out for the evening. After Judith Ann very carefully brought Judy home and walked her to the door, she said, "Do you mind terribly if I kiss you?" She was happy to have the chance after thinking she might never see Judy alive again after Harold and I left Spain with her. They were two very thankful young women.

The phone rang. Jim had finally learned where he would be sent for his new job. "Well, Judy, I guess I'm going to go where it's hot."

"Where, Venezuela?" Judy asked excitedly.

"No, Los Angeles."

"You're kidding!"

"No, I'm not," Jim said. "I have to go in July soon after the fourth."

They both tried to make light of the situation, but they were disappointed with this turn of events. They had hoped he would be sent somewhere closer to home, but they still planned to marry in January.

When the time came for Jim to leave for California, Judy went to his parents' home for the weekend so she could bid him farewell. As Jim was getting ready to leave, he said to Judy, "Now I don't want any tears." Judy held back until she saw his car pull away. Then the tears flowed freely.

I felt such a void when I saw Jim's car drive off. The person I loved and had depended on for so long was not going to be around to lean on any longer. I felt as if my world had lost all purpose and meaning.

In his letters, Jim described how busy he was in Los Angeles. Judy wondered if she would be able to keep up with the fast pace there. Sometimes she became quite depressed. Once she started to sob and said, "Nothing has worked out for Jim and me. We should have been married by now, but everything has turned out wrong for us. He's out in California and I can't be with him. I can't walk right and I'll probably never be able to run and participate in active sports with him. My body is a mess and I won't ever be able to wear a bikini. Why didn't God just let me die? I probably would have been better off."

It was the first time Judy had ever said anything like that. She had been brave through the whole ordeal and had never complained. I could understand her frustrations, but I said, "Don't ever say anything like that again, Judy. We don't know why things are happening as they are, but there is probably a good reason. God brought you back to us, and maybe he has a special plan for you and we just don't understand what it is right now."

A short time later Judy received a letter with a poem by an unknown author enclosed.

> Today upon a bus I saw
> A pretty girl with golden hair.
> I envied her, she seemed so gay.
> I wished that I could be as fair.
> But then, when she arose to leave,
> I saw her hobble down the aisle.
> She had one leg and used a crutch;
> And yet she passed me with a smile.
>
> O God, forgive me when I whine;
> I have two legs, the world is mine.

And then I stopped to buy some sweets.
The lad who sold them had such charm.
I talked with him—he seemed so glad.
If I were late, 'twould do no harm.

And as I left he said to me,
"Please come again, you've been so kind.
It's nice to talk with folks like you
Because, you see," he smiled, "I'm blind."

O God, forgive me when I whine;
I have two eyes, the world is mine.

Then walking down the street I saw
A pretty child with eyes of blue.
She stood and watched the others play.
It seemed she knew not what to do.

I stopped a moment, then I asked,
"Why don't you join the others, Dear?"
She looked ahead without a word,
And then I knew she could not hear.

O God, forgive me when I whine;
I have two ears, the world is mine.

With legs to take me where I go,
With eyes to see the sunset glow,
With ears to hear what I would know . . .

O God, forgive me when I whine;
I'm blessed, the world is mine.

The letter closed with a personal note: "Judy, though you may never know me and though we may never meet, I say to you, 'You're blessed!' Life holds so very much, yet a lot of people overlook life. Take care and God bless you always. Thank God that you are alive and many thanks for reading this letter. W.E.P."

This was a timely letter from another unknown person, another sign that God works in mysterious ways his wonders to perform.

After Judy received this letter I heard no more words of complaint.

23

Judy wanted to take the car out to see if she was still able to drive. I was elected to go with her in our little "Volks." We were both tense. I wondered if her reflexes would be quick enough to stop in an emergency. But she did quite well, even with shifting.

A short time later she said she wanted to try it alone. Both Gene and I anxiously awaited her return. We were relieved that she was away only five minutes. She gained more confidence each time she drove and now drives everywhere by herself.

Judy volunteered to help with the Title I program at one of our schools for six weeks during the summer. It was good therapy for her and kept her occupied in the morning.

She worked until the end of July. Then our family vacationed at my sister Marcia's summer home in northern Wisconsin.

When we returned home Judy saw her orthopedic surgeon in Madison. After a thorough examination the doctor scheduled the hip operation for August 31.

Judy was elated that it would be so soon. She was sure she would have plenty of time to walk correctly before her wedding. She called Jim to share the good news.

An 18-year-old boy in our area had been in a coma for three months following a car accident. I called the mother and asked her if she would like to have Judy and me come and talk to her. She eagerly agreed. When she met us at the door, she gave Judy a big hug and kiss.

We talked for quite a while, encouraging her and answering her questions. A nurse also came and talked to Judy. She asked Judy if she would mind being introduced to all the patients. They'd been praying for the boy since his admittance, and she wanted them to see for themselves that God does answer prayers. Judy was delighted.

Judy was hesitant about going into the young man's room so soon after her ordeal, but she said she would do it. She was shocked—she hadn't realized herself what a coma patient would be like. She couldn't believe she could have been the same way. She said, "I kept wanting to say, 'Wake up, please wake up.'"

His mother said, "Just seeing Judy really did so much to strengthen my faith. You have no idea what it did for me."

August 31   Surgeons removed a piece of calcium the size of a man's fist from the muscle of Judy's left hip. We were told she'd have to stay in the hospital two weeks.

It was a lonely stay. Judy missed Jim's visits and his continued encouragement. It was hard for her to only get a few letters. But she recuperated quickly and was back home in 10 days.

She had to remain on crutches for a month. She went back to the Beloit hospital to continue her physical therapy and to learn to walk correctly all over again. She practiced in front of a large mirror each day. I continued helping her each night with her exercises.

She was unable to jump rope. Her feet just would not lift off the ground. With continued practice she finally was able to get her feet off the ground slightly.

She joined the YMCA so she could work out and swim. One day while Judy was attempting to jump rope, a young woman looked on in amazement. Finally she asked, "Can't you even jump rope?"

"No," Judy replied, "not very well."

She came over to Judy and said, "I'll help you." She took the rope and said, "Now you get in front of me and when I say 'jump,' you jump."

Judy did as she was told because she knew the woman was trying to be helpful, but when the rope came around she was unable to get her feet off the ground.

The young woman shook her head in disbelief. Then she noticed the two very white square areas on Judy's otherwise very tan leg and asked about it. As Judy explained what had happened, the woman's eyes got bigger and her mouth dropped open in awe.

I wish she could know that with continuing determination Judy is now jumping up to 50 times without stopping.

Judy is also jogging around the block each evening. When she first started to run she was very spastic and awkward, but Gene went with her and kept encouraging her. Later when I went jogging with her she passed me and was waiting for me when I finally returned to our starting point.

Judy had learned that she could usually get her way with me if she wanted something or needed sympathy. One day in a burst of tears Janet asked me, "When are you ever going to start treating Judy as an adult instead of a child? She's no baby, you know, so why do you always treat her like one? You give in to her all the time."

"I didn't realize I was doing that," I answered. "I guess after what we've been through with her I was just trying to be good, but I'll have to mend my ways. I'm sorry, Janet. Remind me if I forget."

More than anyone else, Janet made me realize that what I'd been doing was wrong for Judy. She had to learn to rely on herself. I stopped babying her.

Judy wanted to learn to ride her bike, so I started her out as I had when she was a little girl. I ran beside her holding on to the bike until she got her balance. Then I let go and she took off on her own. But when she tried to stop, the bike fell over. She sat in the grass, crying. "Now what would I ever do if I was alone and I fell?"

My first impulse was to help her up and soothe her hurt feelings, but instead I said, "You'd do just as you're going to do right now! Pick up that bike and get back on it."

She looked surprised, but slowly she got up, picked up her bike, got on it, and rode down the sidewalk. She can now ride her bike anywhere.

Whenever Judy says she can't do something, Gene is quick to say, "Well, if you can't do it very well now you'll just have to practice until you do it right." His attitude has helped tremendously in motivating her to try things that seemed impossible.

Judy was busily making wedding arrangements. The wedding dress was bought and the bridesmaids' dresses were ordered and we had taken care of everything except the invitations.

Yet, with all the wedding preparations made, I couldn't make myself believe Judy was getting married. I knew in my heart she was not mentally or physically ready to take on the responsibilities of marriage, but I didn't feel I could say anything. I knew how much she had anticipated getting married even while she was in her coma. It was one of the things that gave her the will to live.

While Judy was in the Beloit hospital, Jim had said, "I want to marry Judy as soon as possible—but only when she is mentally and physically capable of it." I had complete confidence in his judgment. I was sure if I told him Judy was not ready to take on the responsibilities of marriage he would think I was being too protective and possessive. He always seemed as eager as Judy to get married, and I thought he was prepared to take on the responsibilities marriage to Judy would involve. Therefore, in his absence I helped her with the preparations, but I had many doubts in my heart.

Jim was eager for Judy to come out to California to see if she would like Los Angeles and to look for a job and an apartment. Judy also wanted to finalize their wedding plans.

She made reservations for a flight to California in November, just two months after her hip surgery. She had not had sufficient time to recuperate fully from her surgery, but she was eager to see Jim. She left home a very happy person, filled with hope and anticipation.

24

Although I had been trying very hard not to be overly protective, it was difficult for me to put Judy on the bus that would take her to O'Hare, where she would be completely on her own. I remembered the last time I'd said goodbye to her—the next time I saw her she was in a coma. But Judy was very excited and eager. I prayed she would be able to take care of everything by herself when she arrived at the terminal. I asked her to call me after Jim had met her, so I would know she was in good hands.

She arrived in Los Angeles with no difficulty but was two hours late because of a delay at O'Hare. Jim was anxiously awaiting her arrival—and we were anxiously awaiting her call.

Jim hadn't seen Judy for three and a half months. People had written him that Judy was almost like she had been before the accident, so he was expecting to see her in almost perfect condition. As she got off the plane and walked toward him, he could see she had a long way to go.

When we went for a hike up San Gabriel Mountain I bruised my right heel on a stone. I was still limping from my operation on my left hip, and now I was limping on my right foot because of my bruised heel. I had a terrible time getting around, and Jim could see I was not physically ready to take on anything which required much physical strength. It was one more thing gone wrong.

Later Jim said, "I'm so worried about you all day when I'm at work. I'm afraid you might fall in the bathtub or down the stairs. What would I ever say to your parents if anything happened to you when you were out here?"

I could see Jim was having doubts about getting married. I told him I would look for a job in a store if I couldn't find a job as a speech pathologist because I was sure I would have to have my master's degree before I could get one. I did try to find something at a rehabilitation center near Jim's apartment, but they had a freeze on hiring. I knew I would have to find a job where Jim could take me because I couldn't drive in the Los Angeles traffic.

Then Jim said, "I think you should go back to school and get your master's degree so you can get a job in the field you've studied for so long. It would be such a waste if you didn't."

That made me angry. I thought Jim didn't think I was good enough for him if I didn't have my master's. But I now realize how very right he was.

Finally Jim said, "I really don't think you're ready for marriage and I don't think I am either. I don't feel now is the time to get married. Of all the places my company could have sent me, they sent me way out here to Los Angeles. Everything has been working against our marriage from the start. I think we just aren't supposed to get married."

I had been waiting for a call from Judy so I would know how many invitations she and Jim wanted me to order. The call finally came, but Judy had a different message.

"Hi, Mom. I was wondering if you would do something for me?"

"Sure," I answered. "What is it?"

"Will you please call the bridal shop and see if you could cancel the bridesmaids' dresses? Jim and I have decided not to get married."

There seemed to be no tears and there was not a note of sadness in her voice, which surprised me. I was thankful and relieved at her news, but I was sure she must be heartbroken. I told her I would cancel the dresses we had reserved. Then I asked, very concerned, "Are you all right, Judy?"

"Yes, I'm fine," she answered. "I'm having a good time."

I didn't believe her, but I said, "That's good. I'm very glad to hear that."

First I called Erna's, the bridal shop. Erna said she would call the New York designer and let me know. The next day she reported that the designer was very gracious and said they would cancel the gowns without charge, although the seamstress had already started to cut material. I told her how grateful we were to her and the New York designer for understanding our circumstances.

I also canceled the band we had reserved. The leader was a friend of Judy's and said, "No sweat."

Judy called the next day to see if I had been able to cancel arrangements and I told her everything was taken care of. Then I said, "Judy, I want you to know that no one feels bad about this."

"Really?" she asked, very surprised. "Not even Janet?"

"Well, yes, Janet feels bad," I confessed, "but only because she won't be able to get that pretty new dress."

"Yeh, I'll bet," Judy laughed.

I told Judy goodbye and urged her not to feel bad. She said she was sure it was for the best. I anxiously awaited her return home.

What had seemed like a total disaster for Judy turned out to be a blessing. God had planned for Jim to be there when she needed him, but now it was time for her to stand on her own two feet. It was time for her to try to regain her independence and to find herself as an individual. I was thankful that they had both realized marriage at this time would be unwise. I thought they showed mature judgment.

Because Judy is a strong-willed person with deep faith in God, she was determined to pick up the pieces and stop relying on Jim or anyone else. She resolved to make something of the life God had given back to her.

Part of a poem which described her feelings when she was in the hospital again became very meaningful now.

> I know that though he may remove
> The friends on whom I lean,
> 'Tis that I must learn to love
> And trust the one unseen.
> And when at last I see his face
> And know as I am known,
> I will not care how rough the road
> That leads, through Christ, to home.
>
> —Author unknown

And the road was rough as Judy went back to Madison to work for a master's degree in behavioral disabilities.

I found when I studied I had to read out loud before I could comprehend what I was reading. Many times I had to put plugs in my ears to eliminate distracting sounds. But I was determined and I had faith that God would help me.

137

I soon found that I was progressing. I could read and comprehend as well as I did before my accident.

She did her student teaching with three- and four-year-olds who had learning difficulties in a regular classroom. She had to ride the bus with the children five times a month and help lift them on and off the bus when she was not even sure of her own footing in the snow. This age group was hard for her to handle because she was still unable to bend down and move with agility. Her teacher told her any time she felt she couldn't do something she should just tell her, but Judy said she wanted no special consideration. She tried to do everything the other student teachers did.

Each day she had to walk eight blocks to the bus and eight blocks from the bus in bitter cold winter. I worried about her pneumonia recurring, but her health was good.

Judy's favorite Bible verse during this time was Romans 8:28: "And we know that all things work together for good to them that love God" (KJV).

She also read and reread this poem:

> God hath not promised
> Skies always blue
> Flower-strewn pathways
> All our lives through;
> God hath not promised
> Sun without rain
> Joy without sorrow
> Peace without pain.
> But God hath promised
> Strength for the day
> Rest for the labor
> Light for the way
> Grace for all trials
> Help from above
> Unfailing sympathy
> Undying love.

> —Author unknown

Although Judy finished her semester with a 3.7 grade point average, she decided to transfer to the University of Wisconsin, Whitewater, to study for a master's degree in speech pathology.

I wanted to concentrate on speech and language again, specifically language rehabilitation with a focus on aphasic patients. I wanted to work with people on a one-to-one basis. I thought I would be able to relate very well to people who needed rehabilitation.

I am now working part-time as a speech pathologist at an area nursing home and going to the University part-time. I enjoy my work very much. I work mostly with stroke patients. I also enjoy talking with other patients.

In June Judy went back to the University Hospital for a checkup with her orthopedic surgeon. He was pleased with her walk and was happy to know that she was working and continuing her education.

As she was leaving the hospital she met one of her favorite neurologists who had worked with her at the N & R. She spoke to him but he showed no sign of recognition. "Don't you even know who I am?" she asked, smiling.

He recognized the smile and said, "Judy! I can't believe this can really be you. The last time I saw you, you looked like a refugee from Vietnam. I just saw you walking down the hall and I didn't notice any limp, and you were really limping when I last saw you. You're doing just great."

He and Judy talked for half an hour. He had always encouraged her to continue her education.

He explained what had happened to her brain after she was poisoned. "Of course, you know, Judy, in most serious cases of CO poisoning, such as yours, the patient dies."

A short time later on a Milwaukee TV program titled "Life after Life," the reporter asked Judy about her feelings when she was in a coma and so near death. His final question was, "How do you look at life now, Judy?"

This is a summary of Judy's reply.

When I look around at all the beauty and great things there are to life here on earth, I come to an even deeper realization and understanding of the goodness of life. I have a much fuller appreciation of the beauty of nature. There is the blue *of the* sky, *the* green *of the trees, the* white *of the snow. My past experience has brought me closer to the realization of what life is all about.*

My philosophy of life has also changed concerning people. All people are a very important part of life—those close to you and all people *in general. Like the song "People" says, "People who love people are the luckiest people in the world."*

I feel very fortunate that I have my faith in God and that my family had their strong faith during my illness.

When Judy was still comatose I said to Jim one evening, "Do you think this whole ordeal Judy has gone through is going to make her a bitter person when she comes out of the coma?"

"No," he answered thoughtfully, "I think she'll be an even better person than she ever was before."

God has given her a new beginning, a new morning in her life. I believe God has a special plan for her.

Good morning, Judy!